THE FARM TRACTOR

Garrett "Suffolk Punch"

John Appleyard

*Paintings and drawings
by the author*

DAVID & CHARLES
Newton Abbot   London   North Pomfret (Vt)

TO KAY

From too much zeal for what is new
And contempt for what is old,
From putting knowledge before wisdom,
Science before art and cleverness before common sense . . .
Good Lord deliver us.

British Library Cataloguing in Publication Data

Appleyard, John
  The farm tractor.
  1. Farm tractors – History
  I. Title
  629.2′25   TL233

ISBN 0–7153–8876–2

Typeset by ABM Typographics Limited, Hull
Printed in Hong Kong
for David & Charles Publishers plc
Brunel House Newton Abbot Devon

Published in the United States of America
by David & Charles Inc
North Pomfret Vermont 05053 USA

Thanks are due to the following for information and help with the compilation of this book.

J. E. Moffitt, farmer, enthusiast and owner of the National Tractor and Farm Museum, Newton, Stocksfield, Northumberland, who introduced me to the fascinating world of old tractors and kindly lent the paintings for reproduction; Freddie McKendrick, who forged the link; David St John Thomas for encouragement and advice; Charles Cawood; Edwin Bainbridge; Brian D. Thompson; the Bell brothers of Swainby, North Yorkshire; David A. C. Royle; Mr and Mrs R. White of Papamoa, New Zealand; Sir Len Southward of the Southward Museum Trust, New Zealand; Jim Anderson for permission to use his poem; the United States of America Embassy in London; the Canadian High Commission and the National Museums of Canada; the Museum of Lincolnshire Life; the Institute of Agricultural History and Museum of English Rural Life of the University of Reading; editorial staff at David & Charles; Cleveland County Library, Yarm branch; Nadine Peacock for help with the typing of the manuscript, and above all to Les Blackmore, ex-curator of the National Tractor and Farm Museum, who was always ready, willing and able to fill the gaps. The errors and omissions are my own.

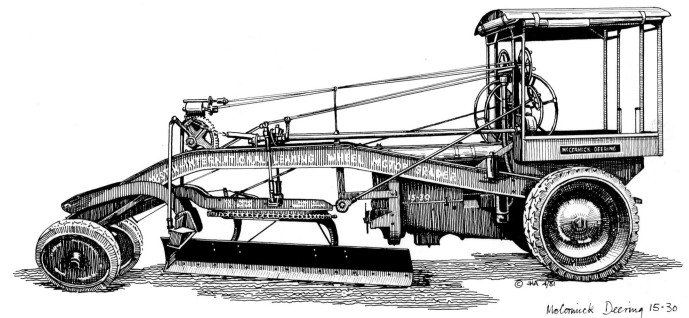

McCormick Deering 15-30
fitted to Austin Motor Grader.

CONTENTS

PREFACE 4

INTRODUCTION 5

THE DEVELOPMENT OF THE FARM TRACTOR 8

DATES IN THE DEVELOPMENT OF THE TRACTOR 12

MUSEUMS WITH ITEMS OF AGRICULTURAL INTEREST 13

THE TRACTORS 16

PREFACE

I am very pleased to pay this tribute to the man who both wrote this book and produced the paintings from which the plates have been made.

Although I am convinced that we all 'make our own luck', it is surprising how much accidental contacts influence our actions. Some years ago I happened to see a prewar Oliver tractor rusting at the roadside near my home farm and finally decided to rescue it from complete disintegration. It cost £8 and was the starting point for the National Tractor and Farm Museum – now comprising a collection of over 100 tractors, as well as other exhibits covering every aspect of farming history.

It is now ten years since I saw a group of large paintings of vintage cars in a restorer's workshop and was impressed by the command of detail. The artist was John Appleyard and I arranged to meet him. At this stage he could possibly distinguish a Ferguson from an Austin Seven but I commissioned him to paint a picture of my Mogul 8/16 tractor to test whether his eye was as good with tractors as with cars. It proved to be – judge for yourself from the plates in this book. The detail was meticulously pre-

cise, the landscape background a convincing complement to the tractor (at the opening of the museum by HM The Queen Mother, The Duke of Northumberland even recognised the actual location of the Mogul painting).

John is enthusiastic about painting, if not obsessive – a characteristic I confess to sharing – my obsessions, among others, being for old machinery and tidiness. So I find myself attracted to the acute observation of detail and the skill to depict it, to which his early background in engineering draughtsmanship no doubt contributes. He has contrived in these paintings to satisfy his own creative urge with few compromises and at the same time to fulfil my requirements for accurate representations of machinery with which I am closely familiar and view with an intensely critical eye. After a lifetime of teaching art, he is now retired and enjoys an ideal regime as his own master, choosing his material from a wide range of subject matter – landscape, old cars, animals, architecture and farm machinery, of course. But, as I know well, being your own boss demands a high level of personal discipline in

organising one's working life. The impression that you are free to please yourself in every respect is false Nor should one allow oneself to be forced into the trap of having time for nothing but work. I consider that John has managed to strike a fine balance between over-obsessiveness and self-indulgence.

His research into the development of the farm tractor has taken a long time, with frequent reference to many sources of original material such as reports of the Royal Agricultural Society and contemporary advertising brochures. The resulting text demonstrates the painstaking care he has brought to the task. His conclusions are his own and the problems of dating tractors and quoting technical details are acknowledged minefields, but I am sure that everyone with an interest in this subject will find much to intrigue him and many an expert will be sent back to his reference books. John's comments upon painting in general and the genesis of each painting will open many eyes and minds to the pleasure reflected in his work.

I trust that this book will bring to John Appleyard the wider recognition which his work merits.

John Moffitt, CBE, FRASE

INTRODUCTION

Why tractors? I have always been interested in mechanical transport – collecting car catalogues (what a pity I hadn't the foresight to hoard them for many are now worth a small fortune), recognising every car on the road at sight. I found a new tram or bus on my route to school exciting, and a highlight of the holiday periods was to cycle into the Plain of York to see Gladiators and Whitley bombers practising what I later came to call fighter affiliation, followed by a visit to Alne station to see the Flying Scotsman – the up and down trains were reputed to pass there, but I can't honestly remember ever seeing this happen. I drew this kind of subject endlessly. Before reaching my teens I'd made a drawing of a design for a forward-control, rear-engined car which I sent off to an engineering exhibition in London – yes, it was shown, although I knew no better than to draw it on lined paper from an exercise book. I now suspect that it owed a lot to a car which appeared in one of those futuristic H. G. Wells-inspired films of the thirties, called *The Tunnel* – memory suggests that it was a Peugeot.

The early years of the war were spent getting very disillusioned about the process of becoming an engineering draughtsman. The slide-rule and log tables (not the calculator in 1940!) dominated the pencil and setsquare and I was relieved to find myself, after long training courses including trips to Canada and the Bahamas, at the navigation table of a Coastal Command Liberator. Aeroplanes did and do fascinate me and although my job did not require me to become involved with the mechanical side of flying, I was in touch with machines and able to appreciate their attraction in relation to nature's display of sea and clouds. I have recently completed a set of aircraft paintings that owes something to those days. My entry into teaching in 1947 set the pattern of my life to the present, for my interest in painting and drawing was reawakened by the opportunities presented by two years at teacher training college and two extremely sympathetic teachers, Miss Hitchman and Mr Dickinson, who opened my eyes to visions whose existence I had not even suspected. Above all, I learned to be curious and to be prepared to be amazed. My last formal contact with any art training had been as a schoolboy, seven to eight years before, and although I'd done some drawing in the RAF, the ability to do quick caricatures or, better still, to hammer out 'Roll out the Barrel' or the 'Warsaw Concerto' on the mess piano seemed more relevant to my companions. At Leeds Training College I really joined the art world and became aware that I knew so little and would never cease to learn. I have always tried to pass on to my students the spirit of Hippocrates' aphorism 'Ars longa, vita brevis', a theme repeated with variations by Chaucer and Longfellow, among others. A further year at Leeds College of Art provided the ideal combination of disciplined learning with freedom to explore the visual world and experiment with materials in expert company. I remember particularly Maurice De Sausmarez, from whom the phrase 'Nice dwawing' (sic) was praise indeed. A few years later on a summer course at Scarborough, I met Harry Thubron, who was very specific about the ways he felt my painting might go (notice the flexibility). The contact with these two influential teachers made me feel that in a small way I had been in on the ground floor of the Basic Design concept which is still the basis of the initial course taken by most art and design students. As I have taken to naming names, I must not neglect the influence of Joe Cole and Owen Yarwood of the then Middlesbrough College of Art, where I spent many hours, both happy and frustrated, in the studios doing my own thing between the end of schooltime and the start of evening classes, with sound advice and help available as needed. Although I was only on the fringe of the college and felt an old man compared with the average full-time student, I was never made to feel an outsider and kept a contact which was very valuable to a new teacher, isolated as the only artist in an academic environment.

I have been lucky to be able to specialise in the teaching of art, at first in a small grammar school in Yorkshire's old North Riding, the start of a lasting appreciation of the beauty of our under-rated North-East region.

My own first car was a Riley Monaco of 1933, acquired not so much out of enthusiasm for the car, but because it was different and I could afford it – just! I came to know it well, out of necessity, but was always more concerned with how it looked than with the under-bonnet mysteries. A blow-back and fire in the carburettor when on a sketching trip was serious to me because it scorched the bonnet paint.

Drawing of cars followed naturally, and when an old car club was formed locally I entered a world of enthusiasts who enjoyed the same things as I did. One day in 1978 I took a telephone call from David Royle, a friend who restores vintage cars professionally and who had commissioned large paintings of some of his restorations – would I contact a man who had a comprehensive collection of old tractors? The man was John Moffitt. In 1964 he had seen a derelict Oliver 70 (page **44**) tractor lying by the roadside near his farm and had considered it a pity that it should be allowed to moulder away and bought it. This was the start of the Hunday Collection, which grew quite rapidly into an important private museum, open to farmers' clubs and such interested groups. My first painting of a tractor, for this museum, was of the International Harvester Mogul 8–16 (page 19), about which I knew nothing but what I could see and understood no more than any mechanically aware person could deduce by looking. I believe that my apprenticeship as an engineering draughtsman helped here and has influenced my style towards precision and the careful recording of detail. But I came to the task with all the arrogant enthusiasm of the new convert and with none of the disaffection of those who know tractors too well from problems of winter starting and long days in an uncomfortable seat. Since then and some forty to fifty tractor paintings later, including a still incomplete set for Hunday, the museum has been opened to the public by HM The Queen Mother and I have learned a good deal about agricultural machinery.

By no means all tractors are ugly, but it is undeniable that they lack the obvious glamour of the veteran Rolls-Royce or the vintage Bentley. There is however often the rugged appeal of a truly functional object, undisguised by stylists and with a natural grace deriving from logical development. In painting, the combination of imagination and craftsmanship can produce a harmony which has nothing to do with the beauty of the subject matter. Art is an attempt to create pleasing forms, but should not be confused with conventional beauty; the pretty and engaging subject is apt to be a stumbling block for the artist.

As tractors can be seen almost anywhere, there are few restrictions to the choice of settings for my paintings. Rural backgrounds provide endless permutations of trees, buildings, hedges, fences and, of course, skies, and although ploughed land appears frequently, I am not obliged to include track and other obligatory pieces of the scene which the painter of railway subjects cannot avoid. Nor am I generally concerned with the geographical or factual anomalies involved in portraying, say, a GWR saddle tank drawing a rake of Pullmans out of King's Cross! But there are some considerations of time and place, as with the crawler tractor (page 31) mentioned later on. Most of the backgrounds (a word I do not like using as it suggests secondary importance – but then the subject *is* really the tractor) are in the north of England, because I live there. I never travel without a sketchbook and camera, and stockpile drawings and shots of odd corners typical of the places where tractors might be parked between jobs or during a break in the working day. Only occasionally am I tempted into using areas of the country which I do not know intimately, such as in the setting for the crawler tractor, a type not much used in the north-east and for which an East Anglian view with ditch and windmill seemed more suitable. In such cases, I find information where I can and trust that I am careful enough in my research to avoid blatant errors. Of all the features found in landscape, trees excite me most with their infinite variations of texture, shape and colour. Singly or in groups, massed in woods or rising alone from a hedge, the tree never fails to attract me and as a vertical element in the dominating horizontals and diagonals of landscape, it is an important ingredient of composition, particularly with a man-made subject in the foreground. I find the softer masses of trees in full leaf more to my taste than the spikiness of winter, but then most of us like to see in pictures what we like to see in reality, and I prefer summer. But there is an intriguing challenge in the play of light on multicoloured and textured trunks and branches against a hard blue winter sky. For some of our harsher northern areas, the austerity of winter trees and less-than-promising skies are more expressive of the character of the land.

No matter what the season and whether my attention is upon landscape or mechanical details, I could not attain real satisfaction without the constantly changing facets of light to puzzle about and struggle with. Some of the paintings show level, grey light, but in most I use strongly directional sunlight to enhance contrasts and to show the structures of the forms I am seeing and the interplay of planes. Many of the notes on individual pictures will be found to dwell upon the use of light to emphasise colour and form in both subject tractor and its setting.

I have written elsewhere of my interest in weather and the visual signs of its origins, and it is impossible to separate the qualities of light which I have used and attempted to portray from the weather conditions at the time. Of course, unlike the photographer, the artist can manipulate these for his own purpose and it should be clear from my analyses of the paintings that the pictures are contrived with a purpose in mind. For example, few of the tractors are shown in the settings where they were seen. Many were drawn and photographed in museums, frequently inside a shed or exhibition hall; others were seen at rallies and in each case, the background was unsuitable as to light or colour, with distracting details of other machinery, figures or particularly the varied clutter of the rally field. So most tractors are painted in places where they may never have been, but hopefully in situations where they could be found and in conditions which make the most of their forms and colour.

Although the youngest of the tractors in these paintings is around forty years old, the condition of most of them is as good as new, usually as the result of a long, knowledgeable restoration, often from an unbelievably dilapidated state. On paintwork of this quality, the effects of light modify the local colour of the object, notably on the tops of bonnets and mudguards where the colder light from the sky subdues red or green. Most painters find the application of paint a sensual, physical experience and there is both tension and excitement in the manipulation of paint to produce the effect of shine by the skilful placing together of contrasting tones. It is also one

of the most gratifying in terms of recognition, for laymen seem to find this technical skill the most amazing 'trick'. I always feel that while it is pleasant to look at paintings and to recognise effects that one has seen for oneself, the best work opens one's eyes to aspects of reality that one has never seen before. I have attempted to avoid the use of highly technical artistic terms in describing the processes by which these paintings were made, hopefully without patronising lay readers. However, if they are to understand an activity in which they do not normally participate, readers must be willing to stretch towards concepts and vision outside their everyday experience and I have flattered them by describing my responses to colour, light and form as the artist understands them. I do not believe that anyone would welcome simplification to the point where one loses the purpose of the explanation.

It really all boils down to what exposition in words makes seem simple – making marks with a brush or other tools in the right colour and tone (ie, lightness or darkness) in the right places. This is not to suggest that all of a painting is carried out *alla prima* (if I might be excused one technicality), that is that all of the marks made are part of the finished upper surface of the painting. In many cases, there is a series of quite complicated layers of paint, some of transparent colour, which build up to the final desired effect. Close to, all of these remain merely controlled marks, but viewed from the correct distance (and don't ask me to define what that distance is!) the total effect should be to produce the image of a thing – tractor, tree, hill or cloud. I am still naive enough (in the best sense of the word) to find pleasure in this process of the laying on of brushmarks and the magic by which they become the illusion of reality.

© JHAP. 10/83
Farmall M, Cub & F12

7

THE DEVELOPMENT OF THE FARM TRACTOR

The word tractor means puller, a neater term than the earlier 'traction engine' from the same etymological source. The prime mover of the modern articulated truck is called a tractor as it pulls trailers. The name was probably first applied to the agricultural machine by the Hart Parr Company of Iowa in about 1902. For the first twenty-five years of its life the tractor was regarded as a mechanical horse and not as a power source in its own right. Implements were drawn by a chain as they would have been by a horse and the tractor tended to be regarded with suspicion and distrust. The reasons for this were mixed and more subtle than just rural conservatism. A smelly, noisy, unreliable machine which probably damaged the soil by its excessive weight was a poor alternative to a familiar, animate team of horses.

Early models

The first internal-combustion-engined tractor was probably the Burger, built in America in 1889, only four years after the first workable motor car driven by an internal-combustion engine to the design of Karl Benz. Like Benz's prototype car, the Burger was a single-cylinder machine, but crudely built and mounted on the chassis of a steam traction engine. Other early models were the Froelich (ancestor of the John Deere) and a tractor by the Case Threshing Machine Company in 1892; in 1894 came the Van Dusen, 1896 the Otto, 1898 the Huber and in 1899 the Morton, which became the forerunner of the International Harvester Company. Another example of the tractor manufacturing giants of the twentieth

century was the Wallis Bear, made by the Wallis Tractor Company who were eventually taken over by Massey-Harris (later Massey-Ferguson).

USA

US developments suffered the drawbacks of anyone who is first in the field with a new design, and the American tractor of the early 1900s was very basic with exposed engine and gearing, heavy at between 5 and 15 tons (5.08 and 15.24 tonnes) and with low horsepower between 15 and 20. By 1914 the needs of a rapidly expanding population and the magnitude of the areas to be worked produced a ready market for the products of almost two hundred tractor manufacturers.

UK

British tractors before 1900 included the Ruston in 1896 and the Saunderson. Dan Albone's Ivel (page 16) of 1902 was the world's first small practical lightweight tractor, particularly suitable for British farms. In 1907 Marshalls of Gainsborough made their first tractor and by 1914 there were fifteen British firms building tractors, although there was only reluctant acceptance of the mechanical horse until the Great War.

1914–18

The Great War was instrumental in forcing the acceptance of mechanisation on British farms. Thousands of horses and men were taken from the land, and without power and with a severely reduced

workforce, farmers realised that the only way to plough and cultivate the land was by using tractors. The situation was aggravated to the point of crisis by the German submarine blockade, but because of the wartime drain on manpower, British firms were incapable of producing the required machinery. Importation of American tractors, such as the International Junior, Mogul and Titan models (pages 19-21), the Waterloo Company's Overtime or Waterloo Boy (page 29) and some models from Case and Wallis, was the only solution if the nation was not to starve. Despite being late into the field, the Fordson Model F (page 37) played a most important role; as well as being the first mass-produced tractor, technically innovative, light in weight and seemingly suicidal in price. About 7000 of the first Fordsons off the production line were exported to Britain.

The twenties

In the depression years of the twenties, undercutting the price of its rivals by as much as 50 per cent, the Fordson thrived and forced many other manufacturers to diversify or become bankrupt.

However, various factors encouraged the development of the internal-combustion-engined tractor at the expense of steam power and from 1925 the paraffin tractor became the standard source of power where tractors were in use. The Great War brought about a boom in oil and an Act of Parliament of 1920, taxing vehicles by weight, completed the decline of steam power which had led the way to mechanised farming.

Tracklayers

Three aspects of the tractor's development require mention before moving into the thirties. First is the tracklayer or crawler type, suitable for work on heavy-land arable farms, where the endless track permits work in conditions unsuitable for wheeled machines and damage to the soil is kept to the minimum. Examples were the Clayton-Shuttleworth (page 31), made between 1916 and 1928, whose tracks left horseshoe-shaped prints on the ground, and the Caterpillar, so well known that the name is commonly incorrectly used as a generic term for the type. It seems that the name Caterpillar was first used at army trials of tracked vehicles in 1908 when a patent by the chief engineer of Richard Hornsby of Grantham was demonstrated. The Great War tank was developed as part of Winston Churchill's 'Landship' project and used much agricultural technology to produce machines to cross the trench systems of the battlefields. The patent of the Roberts track (Roberts was the engineer mentioned above) was bought by Holts of California for £4,000 in 1914 (in 1914 £4,000 = $19,520) after a lukewarm reception by the British Army. It is likely that Holts had already developed their own caterpillar tracks before the purchase of the Roberts/Hornsby patent, but wherever the credit lay, Britain found herself obliged to purchase these rights and a demonstration model when Fosters of Lincoln (rival neighbours of Hornsby) were commissioned to design the machine which became known as the tank. The name derives from the fact that the workforce on the project were led to believe that they were making water tanks for the Middle East. It still seems odd that of two Lincoln-shire companies, both involved in the making of agricultural machinery, the one chosen for the 'Landship' project should be the one with no experience in this field, while the one apparently passed over had already demonstrated such a vehicle to the army. There is a fascinating variety of folklore, probably both apocryphal and libellous, surrounding the situation, including stories of government representatives getting lost and arriving at the wrong factory and of their being generously supplied with liquid hospitality to reduce the likelihood of their discovering the mistake. I believe that it is considered good journalistic practice to print the legend when this proves better than the facts.

Landmarks in the tank's development were the tricycle Killen-Strait, possibly the first armoured tracklayer, and Renault's famous light tank, from which the post-war GP and HI crawler tractors were developed.

Regulations against the use of tracks on the road make this a very specialised type of tractor, replaced in part by the coming of the pneumatic tyre. Crawlers are still necessary on the heavy-land farms of eastern England.

Power take-off

Like steam engines, some early tractors had a belt pulley which could be used to power a machine such as a thresher while the tractor was stationary. By 1918 International Harvester had included a power take-off (PTO) mechanism to power towed equipment when the tractor was in motion. Implements were soon attached direct to the tractor, independent

Holt tractor, as used for artillery towing 1915–18

9

of the road wheels, as on the McCormick-Deering 10–20 (page 35). This type of power take-off usually by shaft, became a standard facility on most tractors.

Three-point linkage
The last development here was the three-point linkage, first experimentally designed about 1920 by Harry Ferguson, but this method of integrating implements with the tractor only fully developed when it was complemented by hydraulics. The basic problem was that lightweight tractors did not have the good traction, or adhesion, required to pull a hard-working implement. Traction could be increased by various methods of ballasting, but this extra weight damaged the soil structure, the problem the lightweight tractor was designed to overcome. The Ferguson three-point linkage used the dynamics of the system's geometry to transform the drag produced by pulling an implement through the ground into weight at the rear wheels, thus improving traction. The Ferguson system is still used by nearly every tractor that is built in the world today.

The thirties
By 1930 the tractor was becoming more established as a part of the farming scene, but the effects of the Depression of the twenties had mostly cancelled out the boost given to tractor production during the Great War. Over the years many changes were to take place, such as the adoption of the pneumatic tyre and the diesel engine; the introduction of hydraulics combined with mounted implements meant that the tractor was no longer a mechanical substitute for the draught horse. While production figures steadily increased, the number of companies involved in the manufacture of tractors declined, often rapidly during the continuing depression of the thirties. Intensive methods of cultivation,

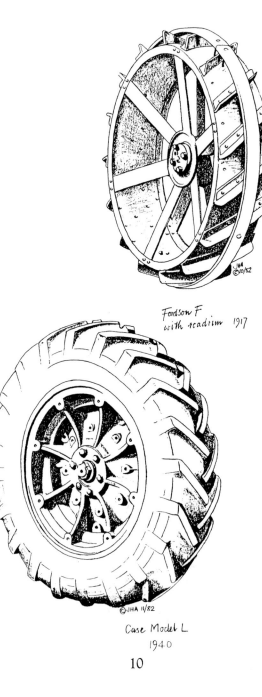

Fordson F
with roadrim 1917

Case Model L
1940

accelerated by the food requirements of a blockaded Britain between 1939 and 1945, brought mechanisation into every aspect of farming.

The pneumatic tyre
The first purpose-made, commercial pneumatic tyre for tractors came from Firestone in 1932 and was first fitted to the Allis-Chalmers Model U (page 50). The problem was to produce wheels with a tread which gave good traction on soft ground, but could be driven to other parts of a farm over other surfaces, including metalled roads. Steel wheels with spuds, spade lugs or cleats were not permitted for use on roads owing to the damage that they caused. In any case, typical rear steel wheels, with about a dozen spade lugs in each of two rows, spaced alternately around the circumference, gave a jerky, spine-jarring progression on hard surfaces. The Fordson (page 47) and some others offered a smooth road rim which could be fitted over the lugs for use on the roads, otherwise lugs had to be removed (which meant forty-eight or more bolts for each rear wheel), allowing the tractor to sit level on its wide rims. Front rims were flanged to give directional stability and more positive steering. But during the later thirties the familiar herring-bone-patterned rear tyre became a regular fitment. Some companies no longer offered a steel-wheel option, although the possibility of continuing on steel, when rubber was scarce between 1939 and 1945, was valuable. Front tyres had little tread, but circumferential ridges had the same effect as the steel flange. If extra weight was needed, water ballast could be carried in the tyres.

Diesel engine
The diesel principle of ignition by compression was not new (the first patents were taken out in 1892/3), and that great automobile pioneer, Karl Benz, had shown it to be practical for tractors in the early

twenties and by 1930 Garretts of Leiston had a diesel tractor in production, which they entered for the World Tractor Trials in opposition to diesel models from McLaren and Blackstone. On the continent, Lanz of Germany brought out the Bulldog (page 53) which had a single-cylinder 2-stroke ignition engine. This was a very reliable tractor that was cheaper to run and maintain than true diesels; this semi-diesel type was taken up by a number of European companies and produced until after the Second World War. However, the supremacy of the spark-ignition petrol/paraffin engine was not challenged until the late forties, despite the appearance of a diesel crawler from Caterpillar in America in 1931 and European models from Deutz, Munktells and Mercedes-Benz in the thirties. Development of the diesel was inhibited in the USA by the low price of petrol, but generally the diesel was cheaper to run as it was a much more efficient engine with its high operating temperatures and compression. For industrial use its disadvantages of weight, bulk, higher price and roughness were outweighed by efficiency and cheap running costs. Starting methods were unorthodox, by blowlamp or cigarette-end, for example, but easier than with some of the temperamental petrol/paraffin engines. Hand cranking a diesel engine was difficult and auxiliary self-starting devices came early.

Implement control
Implements were mounted on the tractor rather than merely towed and at first were lowered or raised by simple mechanical systems, early examples being the John Deere 10/20 General-purpose tractor and the Tillivator attachment for the Fordson. The announcement of a new tractor from Ferguson-Brown Ltd (page 55) in 1939 was more influential than perhaps anyone realised at the time. Conventional in appearance, this tractor incorporated revolutionary ideas of implement integration and control. A hydraulic control lowered and raised the implement and, in later models, maintained a constant draught, irrespective of the tractor's attitude. Ten years later, after a stormy relationship with Ford of America, Ferguson at last secured production of his famous grey TE model (page 57) in Britain, at Coventry, and marketed it as 'The World's Most-copied Tractor'. It cannot be emphasised too often that scarcely a tractor is made anywhere in the world whose implement system does not owe some debt to the Ferguson.

Post-1945
Between 1930 and 1955 the increased reliance upon mechanised methods on the farm, with the added boost of the pressure on farming during six years of war, meant that world tractor production figures soared. However, in a highly competitive situation small companies tend to go to the wall and between the wars many of these disappeared in mergers, takeovers and bankruptcies. In the USA four-fifths of tractor manufacturers failed, yet overall production was up 200 per cent. Several light, two-wheeled tractors in the mould of the self-propelled plough appeared in Britain in the late thirties and a number of small models from firms new to tractor building, like the OTA (page 59) and the Singer Monarch (page 59), were offered postwar. But after 1945 the clearest trend was towards large, multinational companies and American firms such as International Harvester, Massey-Harris (Canada), Allis-Chalmers and Caterpillar expanded to produce in Britain. The B250 was International's first wholly British-designed and made model. By the mid-seventies over 40 per cent of the market was concentrated in the International Harvester Company, Massey-Ferguson and Ford. In the same period, Fiat's figures were in the region of 100,000 units per year.

The fifties saw few fundamental design changes, but greater power was available, especially from the increased use of diesel engines. Larger tractors were required to compensate for reduced rural manpower and the tracked vehicle was one way of applying extra power effectively. In the Second World War the tank was a much more highly developed machine, and after 1945 massive reconstruction projects in civil engineering and building helped to diversify the crawler into the industrial field with tractor-based earth-movers, diggers, dumper-trucks and cranes. Such machines encouraged the development of multi-speed transmission systems (themselves a spin-off from the thriving post-war motor industry) and the mysteries of gear-shifting on the move are already history. The once-revolutionary idea of pivot-steering is commonplace for large tractors and the sound-proof cab is required by law.

A cynical footnote on safety
Since the Second World War, legislation with regard to health and safety at work has increasingly regulated working conditions in commerce, trade and industry in the UK. During the fifties and sixties many tractors sprouted roll-bars, often incorporated into home-made cabs of hardboard and canvas. It may be that the drive for increased home food production pushed the tractor on to land outside the obvious arable areas, with the added dangers of cultivating steep, rough fields. However, it was not until 1974 that crash canopies were required by law and even then it was argued that the estimated cost of £4 million per annum to save forty lives was too high (see contemporary Parliamentary debates). Other industries apparently offered a better return than one life per £100,000 of expenditure on safety.

DATES IN THE DEVELOPMENT OF THE TRACTOR

(Tractors illustrated are in italics)
1770 Edgeworth's Endless track
1838 Foundation of the English Agricultural Society
1842 J. I. Case Company founded
1846 James Boydell's Endless Railway & Footed Wheel
1847 Barrat Frere's Pick Axer
1854 Fowler's first design for a steam ploughing engine
1855 Bark's tracklayer
1875 Aveling engines used by French to haul artillery
1876 Marshall & Sons founded; Russians buy Aveling & Fowler engines as artillery tugs
1878 Fowler engines used to haul artillery by Germans in Franco-Prussian War
1882 Guillaume Fender's All-round track
1886 Applegarth's steam tracklayer
1888 Batten (US)
1889 Burger (US); John Charter's internal-combustion-engined tractor
1891 Massey-Harris merger
1892 Froehlich (US); Case (US); Petter (UK); Scott (UK); the first diesel patents
1894 Van Dusen (US)
1896 Otto; Ruston (UK); Hornsby-Ackroyd (UK)
1897 Cougis (France)
1898 Huber; Saunderson (UK); Marshall (UK)
1899 Morton; steam engines used in S. African war
1900 Scott's motor cultivator
1902 Formation of International Harvester Company in Chicago; Hart-Parr; Wallis Bear; *Ivel to 1916*; Petter's self-propelled portable engine
1903 Petter's Intrepid; Ransomes; Drake & Fletcher
1904 Holt's No 77 crawler
1906 Saunderson Universal silver award at Derby; Dan Albone died
1907 Marshall 30hp export model; Deutz
1908 Holt's tracklayer; Hornsby tracklayer demonstrated to the army; Winnipeg trials to 1914
1909 Wallis & Stevens lightweight steam tractor; Hornsbys win War Office trial for cross-country tracklayers
1910 First official tests by RASE at Baldock; Rumely 'Kerosene Annie'

1911 Wyles motor plough to 1921; first internal-combustion-engined tractors from Case; Daimler
1912 Ideal; Waterloo Boy; Parrett
1913 *IH Twin 12–25 to 1918*; Wallis Cub; Fowler internal-combustion-engined cable plough; Walsh & Clark oil cable plough; Munktells
1914 Roberts' track patent sold to Holts (US); *Crawley motor plough to 1924*; *IH Mogul 8–16 to 1917*; *Weeks to 1925*; Foster halftrack
1915 *IH Titan 10–20 to 1922*; *Wallis Cub Junior to 1919*; *Walsh & Clark Victoria to 1922*; Petter's Iron Horse; Garrett Suffolk Punch; Foster's Little Willie successfully tested
1916 *Alldays & Onions Mk 1 to 1919*; *Clayton-Shuttleworth to 1928*; *Saunderson model G to c 1931*; electric starting from Moline; *Waterloo Boy (US)/Overtime Model N (UK) to 1923*
1917 Emerson Brantingham 12/20 to 1928; *Fordson F to 1928*; *Glasgow to 1920*; *IH Junior to 1922*; official foundation of Women's Land Army; Ford Cork factory opened
1918 Fiat 702; Little Giant to 1927; Moline Universal C & D to 1923; McLaren motor windlass; Summerscales (steam); IH introduce power take-off; foundation of Agriculture and General Engineering company; Waterloo company became John Deere
1919 Austin R to 1927; Peterbro 18/35 to 1930; Twin City 12/20 to 1929; Pick; Martin; Walter Christie's first tank; Fordson F made in Northern Ireland to 1922; SMMT trials at South Carlton; Renault crawlers; Rumely Oil Pull Models to 1931
1920 British Wallis to 1929; Blackstone; Santler & Waller one-way motor ploughs; Ferguson experiments with three-point linkage; Lincoln trials
1921 IH, 14–30
1922 Eagle F16/30; Fowler ploughing engine to 1929; *McCormick-Deering 10/20 to 1939*; Howard prototype rotary hoe
1924 Kirov, the first Soviet tractor; Sentinel (steam); Farmall series started production
1925 Case 18/32 to 1929; Ronaldson & Tippett (Australia); Caterpillar Tractor company founded; Saunderson taken over by Crossley to 1932; Vickers Aussie to 1933; McDowall's electric tractor; Landini
1926 Morris tracklayer (based upon Morris-Martell tank); Armstrong-Siddeley four-wheel-drive articulated tractor (never produced)
1927 *Fowler Gyrotiller to 1939*; Deutz diesel; Cassani diesel
1928 *Howard (Australia) to 1936*; Patisson to 1932; McLaren diesel; Rushton; United tractor (later Allis-Chalmers

Model U); Wallis joined Case (Massey-Harris); Fendt
1929 *Case L to 1940*; Case C to 1939; Caterpillar 10 to 1931; *Fordson N to 1945*; Garrett diesel; Minneapolis-Moline company founded; United tractor consortium; Oliver/Hart-Parr merger
1930 Caterpillar 22; Cletrac EG 31; Latil KTL; *Massey-Harris 4×4 to 1936*; Marshall two-stroke diesel; Wallingford trials; new road traffic acts imposed axle road restrictions; Marshall take over Clayton Shuttleworth
1931 Caterpillar diesel; Foden (steam); Garrett diesel broke world non-stop ploughing record at 977 hours; Farmall F30; Allis-Chalmers take over Rumeley
1932 Fowler Gyrolette to 1937; *IH F12 to 1938*; IH F20 to 1939; IH W30 to 1940; introduction of Firestone low-pressure pneumatic tyre
1933 *Allis-Chalmers Model U to 1950*, reaches 64.28 miles/hour (103.4km/hour); Ferguson's Black Tractor; Massey-Harris 25/04; Fordson tractor production transferred to Dagenham
1934 *John Deere Model A to 1952*; Twin City KTA to 1938
1935 Caterpillar R2; Caterpillar D6 to 1959; *Lanz Bulldog*; *Oliver 70 to 1948*
1936 Case RC to 1939; *Ferguson-Brown to 1939*; John Deere BO to 1947; Marshall M to 1945; *Ransome's MG series to 1959*
1937 Allis-Chalmers WF to 1941; Porsche 110 (Volksschlepper); John Deere L to 1940; formation of Ferguson-Brown company; Farmall series changed from grey to red
1938 Caterpillar D2 to 1955; IH F14 to 1939; IH W14 to 1939; IH Farmall A to 1947; Ferguson-Brown demonstrated to Ford; Fordson colour changed from blue to orange
1939 *Case R*; Case tractors changed from grey to orange; Case DC to 1955; Cletrac HG 42 to 1951; Ford-Ferguson 9N to 1947; Ferguson/David Brown partnership ends; David Brown VAK I; IH Farmall H to 1953; 100th Royal Show at Windsor
1940 Case SC to 1955; IH W4 to 1952; Massey-Harris 102 Junior; Allis-Chalmers diesel crawlers, models H7, 10 & 14; Fordsons now painted green
1941 IH W9 to 1951; Porsche/VW Ostradschlepper
1943 Avery R to 1951
1944 Oliver/Cletrac merger
1945 Fordson E27N to 1952; David Brown VAK IA to 1947; Marshall Series 1 to 1947; Bristol crawler to 1948; Kendal PU8 to 1950; Loyd
1946 Bean Toolframe
1947 *Ferguson TE 20 to 1956*; Fordson 8N to 1952; *David Brown*

Cropmaster/Taskmaster to 1953; Nuffield M4; Allis-Chalmers B to 1955; IH Farmall Cub to 1964; Allgaier; Massey-Harris 744D to 1954; Massey-Harris started production in UK; Allis-Chalmers started production in UK; Marshall Series II to 1949

1949 IH BM production started in UK; Fiat crawler to 1951; Garner to 1955; Turner Mk2B to 1951; OTA; Newman; Byron; Glave; *Marshall Series III to 1952*

1950 Lanz Bulldog BH–10–1; Volvo T–25; Fraser diesel prototype; Caterpillar's first overseas plant, in UK; *Trusty Steed four-wheel tractor*

1951 BMB President to 1956; Cuthbertson Water Buffalo

1952 Ferguson receives $9.25 million damages from Ford; *Singer Monarch to 1956;* Vickers VR–180 crawler to 1959; Ford Diesel Major; Howard Platypus crawler

1953 Massey-Ferguson amalgamation; Lanz full diesel

1954 Roadless RT20 crawler; County Four Drive

1955 Vickers Vigor; David Brown 2D Tool carrier to 1961

1956 IH B250, first British-made and designed tractor from IH; David Brown DI

1957 Mannesman; Doe Triple D; Massey-Ferguson MF35 replaces FE35 in MF red

1958 Bukh D30 (Denmark) to 1959; Ford lightweight Dexta

1959 Massey-Ferguson takes over Perkins

1960 Harry Ferguson dies; White Motor company bought out Oliver

1963 Nuffield tractor production moved to Bathgate; Renault took over Mannesman

1964 Fordson production moved to Basildon; County 'Sea Horse' amphibian

1965 Massey-Ferguson DX range launched; David Brown colour changed from red to white

1967 Nuffield new-look tractor; Tenneco Inc take over Case

1968 BMC/Leyland merger

1970 Start of NIAE development of cheap Third World tractor; Kubota

1972 Tenneco Inc take over David Brown

1974 Crash canopies required by law (UK)

1977 Industrial Engine Sales Buffalo for Third World use

1978 JCB 520 introduced for agricultural work; Russell 3D tool carrier

1979 Leyland interest in crawlers sold to Aveling-Marshall; Tranter GP tractor

---

MUSEUMS WITH ITEMS OF AGRICULTURAL INTEREST

UK and Eire

AVON: Bristol Industrial Museum, Princes' Wharf, Bristol 2; Oakhill Manor Museum, Oakhill, Nr Bath (transport models)

BERKSHIRE: Museum of English Rural Life, Whiteknight's Park, Reading

BIRMINGHAM: Museum of Science & Industry, Newhall St, Birmingham 3 1RZ

BUCKS: Gallery of Bucks Rural Life, Church St, Aylesbury; Stacey Hill Collection of Industry & Rural Life, Southern Way, Wolverton, Milton Keynes; West Wycombe Motor Museum, Chorley Rd, W. Wycombe (stationary engines)

CAMBS: Farm Land Museum, 50 High St, Haddenham, CB6 3XB; George Brittain's Collection, Georgina Villa, Hail Weston, Huntingdon

CLEVELAND: Kirkleatham Old Hall Museum, Redcar (tractors); Newham Grange Leisure Farm, Coulby Newham, Middlesbrough

CORNWALL: Camborne Carriage Collection, Lower Grilles Farm, Treskillard (waggons); Farm Museum, Mawla Well, Mile Hill, Redruth (steam engines); Lanreath Mill & Museum, Lanreath, Looe (tractors); North Cornwall Museum & Gallery, The Cleave, Camelford (waggons)

CUMBRIA: Abbot Hall Museum of Lakeland Life & Industry, Kendal (waggons); Levens Hall Steam Collection, Levens (steam engines)

DERBYS: Elvaston Castle, Borrowash Rd, Elvaston (horse vehicles); Riber Castle Fauna Reserve & Wild Life Park, Matlock (tractors)

DEVON: Alscott Farm Museum, Shebbear (tractors); Buckfastleigh Farm Museum, Dial Court, Buckfastleigh (tractors); Furze Farm Park, Bridgerule, Holsworthy; James Countryside Museum, Bicton Park, Budleigh Salterton (tractors); Riverton Farm, Staverton, Totnes; Steam & Countryside Museum, Exmouth; Sticklepath Museum of Rural Industry, Okehampton

DORSET: Army Tank Museum, Bovington Camp, Wareham, BH20 6JG (early tracked vehicles); Country Life & Farm Museum, Charmouth (tractors); Transport & Rural Museum, Bournemouth

DURHAM: Beamish North of England Open-air Museum, Nr Chester-le-Street

EIRE: Irish Agricultural Museum, Wexford

ESSEX: Cater Museum, Billericay; Cottage Museum, Great Bardfield; Upminster Tithe Barn Agricultural & Folk Museum, Upminster

GLOUCS: Cotswold Countryside Collection, Fosseway, Northleach (waggons); Folk Museum, 99–103 Westgate St, Gloucester; R. A. Lister Company Museum, Dursley (machinery); Smerrill Farm Museum, Kemble, Nr Cirencester; Stroud & District Museum, Stroud

HAMPSHIRE: Breamore Countryside & Carriage Museum (tractors); Hampshire Farm Museum, Manor Farm, Botley (steam); Hollycombe Steam Collection, Hollycombe House, Liphook (steam engines); National Motor Museum, Beaulieu (tractors)

HEREFORD & WORCS: Country Centre, Clowstop, Kidderminster; Hereford & Worcestershire County Museum, Hartlebury Castle, Kidderminster (waggons); Kirbee Rural Crafts Museum, Whitchurch, Ross on Wye (waggons)

HERTS: Pitstone Local History Society Museum, Tring; St Albans City Museum, Hatfield Rd, St Albans (waggons); Weston Country Museum, Weston, Nr Hitchin

ISLE OF MAN: Gilbey Horses, Ballacallin Bee, Marown (waggons); The Grove Rural Life Museum, Nr Ramsey

ISLE OF WIGHT: Albany Steam & Industrial Museum, Newport; Yafford Mill & Farm Museum, Shorwell (tractors)

JERSEY: The Jersey Museum, Hougue Bie, Five Oaks

KENT: Agricultural Museum, Wye College, Court Lodge Farm, Brook, Ashford (tractors)

LANCS: Tom Varley's Museum of Steam, Gisburn, Nr Clitheroe; Witton Park Visitor Centre, Preston Old Rd, Blackburn

LEICS: Leicestershire Museum of Technology, Abbey Pumping Station, Corporation Rd, Leicester; Rutland County Museum, Catmos St, Oakham (waggons)

LINCS: Church Farm Museum, Skegness (waggons); F. A. Smith Collection – see Wroughton, Wiltshire; Mawthorpe Collection of Bygones, Alford (tractors); Museum of Lincolnshire Life, Burton Rd, Lincoln (waggons)

LONDON: Knightscote Farm Museum, Breakspear Rd, Harefield; Science Museum, Exhibition Rd, South Kensington (tractors)

NORFOLK: Bygones at Holkham, Holkham Park (tractors); Chalk's Carriage Museum, Barrack St, Inner Ring Rd, Norwich (waggons); Norfolk Rural Life Museum, Beech House, Gressenhall, Dereham; Thursford Museum, Nr Fakenham (steam engines)

NORTHANTS: Burt Family Private Collection of Waggons & Equipment (waggons); Naseby Battle & Farm Museum, Purlieu Farm, Naseby (tractors)

N. IRELAND: Ardress Agricultural Museum, Co Armagh; Ulster Folk & Transport Museum, Cultra Manor, Holywood, Co Down and Witham St, Newtonards Rd, Belfast (waggons)

NORTHUMBERLAND: Eyemouth Museum, Auld Kirk, Market Place, Berwick upon Tweed; Hunday National Tractor & Farm Museum, Newton, Stocksfield (tractors)

NOTTS: Industrial Museum, Wollaton Park, Nottingham

OXON: Cotswold Folk & Agricultural Museum, Asthall, Nr Burford; King William, Ipsden (a public house display) (waggons); Manor Farm Museum, Cogges, Nr Witney

SCOTLAND: Angus Folk Museum, Kirkwynd Cottages, Glamis, Tayside; Glasgow Museum of Transport (steam engines); Glencoe & North Lorn Folk Museum, Glencoe, Highland; Glenesk Museum, The Retreat, Glenesk, Grampian; Hamilton District Museum, 129 Muir St, Hamilton, Strathclyde; Highland Folk Museum, Kingussie, Highland; Jim Russell, Pilmuir, Lundin Links; Myreton Motor Museum, Aberlady, Lothian (tractors); Royal Scottish Museum, Chambers St, Edinburgh, EH1 1JF

SHROPSHIRE: Acton Scott Working Farm Museum, Nr Church Stretton (waggons); White House Museum of Buildings & Country Life

SOMERSET: Brympton D'Evercy Country Life Museum, Yeovil; Somerset Rural Life Museum, Abbey Farm, Chilkwell St, Glastonbury (waggons/machinery)

STAFFS: Park Farm, Grangewood, Burton on Trent; Staffordshire County Museum, Shugborough

SUFFOLK: East Anglia Transport Museum, Carlton Colville, Lowestoft; Easton Park Farm & Motorcycle Collection, Nr Whickham Market; Museum of East Anglian Life, Stowmarket (waggons); Old Mill House, Framsden, Stowmarket; Sotterley Agricultural Museum, Alexander Farm, Sotterley, Beccles (tractors)

SURREY: Garelock, Elstead Rd, Milford (waggons); Old Kiln Agricultural Museum, Reeds Rd, Tilford, Farnham (tractors)

SUSSEX: Drusillas, Berwick, Nr Alfriston (waggons); Michelham Priory, Upper Dicker (waggons); Norton's Farm Museum & Farm Trail, Battle; Tithe Barn Farm, Chuck Hatch, Hartfield (tractors); Valelands Schools' Farm & Museum, Marle Green, Horam; Wilmington Museum, Wilmington Priory

TYNE & WEAR: Newcastle Science Museum, Blandford House, West Blandford St

WARKS: Mary Arden's House, Wilmcote

W. MIDLANDS: See BIRMINGHAM

WALES: Erdigg, Wrexham, Clwyd; Gwent Rural Life Museum, The Malt Barn, New Market St, Usk, Gwent; Model Farm Folk Collection, Wolvesnewton, Chepstow, Gwent; Powysland Museum, Salop Rd, Welshpool; St Fagan's Welsh Folk Museum, Cardiff, S. Glamorgan; Welsh Industrial & Maritime Museum, Cardiff, S. Glamorgan (steam)

WILTSHIRE: Castle Farm Folk Museum, Marshfield, Nr Chippenham; Coate Agricultural Museum, Swindon (waggons); Great Barn Folk Life Museum, Avebury; Lackham Agricultural Museum, Lackham College of Agriculture, Lacock, Chippenham (tractors); Shaw Agricultural Museum, Swindon; Wroughton, Swindon: Science Museum store, including F. A. Smith collection (tractors)

YORKSHIRE: Ambrose Brennand Collection, Nr Ingleton (B6255) (tractors); Bradford Industrial Museum, Moorside Mills, Moorside Rd, Eccleshill, West Yorks; Ryburn (Pennine) Farm Museum, Ripponden, West Yorks; Ryedale Folk Museum, Hutton-le-Hole, North Yorks; Top Farm Museum, West Hardwick, Nr Nostell, Wakefield, West Yorks (tractors); Worsbrough Mill Museum, Barnsley, South Yorks; Yorkshire Farm Machinery Preservation Society Museum, Murton Livestock Centre, York (tractors)

REFERENCES: Toulson, Shirley. *Discovering Farm Museums & Farm Parks* (Shire, 1977).

Historical Transport, Annual.

Alcock, S. (ed). *Museums & Galleries in Great Britain & Ireland*, Annual. Pub. Business Press International.

Deith (ed). *Steam Yearbook & Preserved Transport Guide*, Annual. Tee pubs.

Hudson and Nicholls. *The Directory of Museums* (MacMillan).

Hudson (ed). *The Good Museums Guide* (MacMillan).

USA

ARKANSAS: Arkansas County Agricultural Museum, Stuttgart; Plantation Museum, Scott (tractors); Robinson Farm Museum & Heritage Centre, Everton (machinery)

CALIFORNIA: San Joaquin County Historical Museum, Lodi (tractors); Tulare County Museum, Visalia

COLORADO: Forney Transportation Museum, Denver (tractors)

DELAWARE: Delaware Agricultural Museum, Dover (tractors)

DC: Agricultural History Society Museum, Washington

GEORGIA: State Museum of Agriculture, Tifton

HAWAII: Grove Farm Homestead, Lihue

IDAHO: Van Sylke Museum, Caldwell

ILLINOIS: Clayville Rural Life Centre & Museum, Pleasant Plains; Deere & Co Administrative Centre Museum, Moline (tractors); John Deere Historic Site, Dixon; Museum of Science & Industry, Chicago; Stone Mill Museum, Sandwich (machinery)

INDIANA: Discovery Hall, South Bend

IOWA: Mills County Historical Society & Museum, Glenwood (machinery); Museum of Midwest Old Settlers' & Threshers' Association, Mount Pleasant (tractors); Nelson Pioneer Farm & Crafts Museum, Oskaloosa

KANSAS: Agricultural Hall of Fame & National Centre, Bonner Springs (machinery)

LOUISIANA: Bayou Fort Museum, Cloutierville

MAINE: Fryeburg Fair Museum, Fryeburg; Matthews Farm Museum, Union

MARYLAND: Carroll County Farm Museum, Westminster; National Colonial Farm, Accokeek; Oxon Hill Farm, Oxon Hill

MASSACHUSETTS: Drumlin Farm Education Centre, Lincoln; Hadley Farm Museum Centre, Hadley

MICHIGAN: Greenfield Village & Henry Ford Museum, Dearborn (tractors); Montague Museum & Historical Society, Montague; Museum of West Central Michigan Historical Society, Lowell; Waterloo Farm Museum, Stockbridge

MINNESOTA: Gibbs Farm Museum, St Paul; Hemp Old Vehicle Museum, Rochester (tractors); Marshall County Historical Society Museum, Warren (machinery); Redwood County Historical Society, Redwood Falls (machinery); Renville County Historical Society Museum, Morton (machinery)

MISSISSIPPI: Patee House Museum, St Joseph

MONTANA: Central Montana Agricultural Research, Moccasin; Museum of N. Eastern Montana Threshers' & Antique Association, Culbertson (tractors)

NEBRASKA: Saline County Historical Society, Dorchester (machinery); Stuhr Museum of the Prairie Pioneer, Grand Island (machinery)

NEW HAMPSHIRE: New Hampshire Farm Museum, Milton

NEW JERSEY: Longstreet Farm, Holmdel; Cook College of Agriculture Museum, New Brunswick; Space Farms Museum, Sussex (machinery)

NEW YORK: Agricultural Museum, Stone Mills; Cayuga County Agricultural Museum, Auburn; Daniel Parrish Witter Agricultural Museum, Syracuse (vehicles); Farmers' Museum, Cooperstown (vehicles); Mulford Farm Complex, East Hampton; Museum of New York Agricultural History Society, LaFargeville; Queens County Farm Museum, Floral Park; Shingle House, Warwick (machinery)

NORTH CAROLINA: Pick Shin Farm Living Farm, Dobson
NORTH DAKOTA: Lansford Threshers' & Historical Association, Lansford; Makoti Threshers' Museum, Makoti
OHIO: Aullwood Audubon Farm, Dayton; Clark County Historical Society Museum, Springfield; Durell Farm Museum, Columbus; Malabar Farm State Park, Lucas; Rock Mill Farm Museum, Lancaster; Sauder Museum Farm & Craft Village, Archbold; Wood County Infirmary (Farm Museum), Bowling Green
PENNSYLVANIA: Pennsylvania Farm Museum of Landis Valley, Lancaster (vehicles); Quiet Valley Living Historical Farm, Stroudsburg
RHODE ISLAND: Museum of Portsmouth Historical Society, Portsmouth
SOUTH CAROLINA: Neeses Farm Museum, Neeses
SOUTH DAKOTA: Museum of South Dakota Agriculture, Brookings (machinery); Pioneer & Automobile Museum, Murdo (tractors)
TENNESSEE: Tennessee Agricultural Museum, Nashville; Tipton-Haynes Living Historical Farm, Johnson City
TEXAS: City County Pioneer Museum, Sweetwater; Fort Bend County Museum, Richmond; Hal S. Smith Farm Machinery Museum, Cresson (tractors); White Deer Land Museum, Pampa (machinery)
UTAH: Man & His Bread Museum, Wellsville (tractors); Wheeler Historic Farm, Salt Lake City
VERMONT: Walker Museum, Fairlee (vehicles)
VIRGINIA: Belle Grove, Middletown; Cyrus H. McCormick Memorial Museum, Steele's Tavern (machinery); Humpback Rocks Visitor Centre, Waynesboro
WASHINGTON: Fort Walla Walla Museum, Walla Walla
WEST VIRGINIA: Watters Smith Memorial State Park, Lost Creek (machinery)
WISCONSIN: Calumet County Historical Society, Chilton; Galloway House & Village, Fond Du Lac; Stonefield Village & State Farm Museum, Cassville (machinery); The Farm, Sturgeon Bay

CANADA

ALBERTA: Centennial Library & Museum, High Prairie; Garden Park Farm Museum, Alberta Beach (machinery); Hanna Pioneer Museum, Hanna (machinery); Historical Village, Innisfail (machinery); Homestead Antique Museum, Drumheller; Reynolds Museum, Wetaskiwin (900 tractors); Sir John A. McDonald Museum, Carben (machinery)

BRITISH COLUMBIA: British Columbia Forest Museum, Duncan (tractors); Central Saanich Pioneer Museum, Saanichton (machinery); Farm Machinery Museum, Fort Langley (tractors); Father Pandosy Museum, Kelowna (machinery); Horsefly Historical Museum, Horsefly; Lilloet Museum, Lilloet (machinery)
MANITOBA: Cook's Creek Heritage Museum, Cook's Creek (machinery); Crossley Museum, Grandview (machinery); Hamiota Pioneer Club Museum, Hamiota (machinery); Manitoba Agricultural Museum, Austin (tractors); Manitoba Automobile Museum, Elkhorn (machinery); Pembina Threshermen's Museum, Winkler (machinery); Woodlands Pioneer Museum, Woodlands (machinery)
NEW BRUNSWICK: Automobile Museum, St Jacques (tractors); Lutz Mountain Meeting House, Moncton
NOVA SCOTIA: Nova Scotia Museum, Halifax
ONTARIO: Backus Mill & Agricultural Museum, Port Rowan; Champlain Trail Museum, Pembroke; Doon Pioneer Village, Kitchener; Founders' Museum, Fort William; Glengarry Pioneer Museum, Dunvegan (machinery); Halton Region Museum, Milton; Kitty Historical Museum, Frankville; Madonna House Pioneer Museum, Combermere; National Museum of Science & Technology, Ottawa; Norwich Pioneers' Society Museum, Norwich; Ontario Agricultural Museum, Milton (machinery); Spruce Lane Farm, Oakville; Tilbury West Agricultural Museum, Comber (machinery); Wellington County Museum, Elora
PRINCE EDWARD ISLAND: Car Life Museum, Benshaw (machinery); Farm Development Museum, Freeland (machinery); Garden of the Gulf Museum, Montague; Old Mill Museum, Long River
QUEBEC: Compton County Historical Museum, Eaton
SASKATCHEWAN: Antique Tractor Museum, Maple Creek (tractors); Carscaden's Museum, Plenty (machinery); Earl Grey Centennial Museum, Earl Grey (machinery); Frobisher Threshermen's Museum, Frobisher (tractors); Naicam Museum, Naicam; Old Homestead Museum, Herbert; Our Heritage, Star City (machinery); Pioneer Village Museum, Melfort; Rocanville & District Museum, Rocanville (tractors); Thompson Museum, Readlyn (tractors); Western Development Museum, Yorktown & Saskatoon (tractors)

AUSTRALIA

CANBERRA: Billabong Park Horse Era Museum, Stirling Avenue, Watson (North), Canberra City
NEW SOUTH WALES: Armidale Folk Museum, Armidale; Folk Museum, Mcquairie House, 1 George Street, Bathurst; Folk Museum, High Street, Bowraville; Historic Museum, School of Arts building, Dowling Street, Dungog; Museum of Historic Engines, Marsden Bridge, Crookwell Road, Goulburn; Historical Museum, Homer Street, Gundagai; Richmond River Historical Society Museum, 135 Molesworth Street, Lismore; Historical Museum, Millthorpe, Nr Orange; Experiment Farm Cottage, 9 Ruse Street, Parramatta; Historical Museum, Court House, Singleton; Museum of Applied Arts and Sciences, Harris Street, Broadway, Sydney; Wellington Museum, Warne and Percy Streets, Wellington; Folk Museum and Arts Centre, Wentworth; Authenticated Australian Settlers Village, Wilberforce
QUEENSLAND: Early Street, 75 McIlwraith Avenue, Norman Park, Brisbane; Redlands Museum, Cleveland Showgrounds, Brisbane; Castle of Yesteryear, Nerang Road, Broadbeach, Gold Coast; Vintage and Veteran Car Museum, Coolangatta, Gold Coast (tractors); Dalby Museum, 22 Drayton Street, Dalby, Darling Downs
SOUTH AUSTRALIA: Folk Museum, Burra; National Trust Museum, Matta House, Kadina, York Peninsula; Old Mill Museum, Mill Hill, Port Lincoln (working models); Barossa Valley Museum, 47 Murray Street, Tanunda
TASMANIA: Good Old Days Folk Museum, Swansea
VICTORIA: El Dorado, El Dorado; Agricultural Museum, Jeparit; Kyneton Historical Museum, Piper Street, Kyneton; Folk Museum, High Street, Maldon; Swan Hill Folk Museum, Horseshoe Bend, Swan Hill; Wilson's Promontory Museum, Wilson's Promontory
WESTERN AUSTRALIA: The Old Farm, Strawberry Hill, Albany

NEW ZEALAND

AUCKLAND: Museum of Transport and Technology, Western Springs, Auckland; Matawhero Museum, Matawhero
CANTERBURY: Langlois-Eteveneaux House, Akaroa; The Ferrymead Trust, Bridle Path Road, Heathcote, Christchurch; Yaldhurst Transport Museum, School Road, Christchurch; Geraldine Farm Machinery Museum, Geraldine (tractors); Waimate Historical Museum, Harris Street, Waimate
OTAGO: Lakes District Centennial Museum, Buckingham Street, Arrowtown; Taieri Historical Park, Outram
WELLINGTON: Southward Museum Trust, Otaihanga Road, Paraparaumu; Tokomaru Steam Engine Museum, Tokomaru

THE IVEL · 1903

By modern standards, when most tractors are the sophisticated products of large international conglomerates, a three-wheeled vehicle designed by a racing cyclist, who was also a cycle builder, does not seem to be the formula for a world-beater. This is especially apparent when one considers that the opposition was primarily from the USA, where the incentives for development were greater and the industrial processes required were being rapidly expanded.

The designer was Daniel Albone and his Ivel tractor appeared in 1902 (named after the Bedfordshire river) after five years of experiments.

From a rural background, Albone was very aware of agricultural machinery. To be successful, early tractor designs had to persuade the farmer that they were capable of better performance than the horse team or the steam ploughing set. Cost, reliability and weight had to be right to overcome apathy and opposition. Despite a general lack of interest from farmers, the model remained in production for 14 years, winning a Royal Agricultural Society's silver medal in 1904 and 29 other awards by 1906. The firm was sound enough to attract the support of the distinguished motorists S. F. Edge and Charles Jarrott, who became directors of Ivel Agricultural Motors Limited. Exports of this potential world-beater were made to 30 countries including the USA, but the driving force of the company was lost when Dan Albone died in 1906.

With its bonnet over the engine (not shown) the Ivel was quite modern in layout but had most of the design features of very early models – girder chassis, tank water cooling 30gal (136 litres) with no radiator, cable steering via a single, flanged front wheel and chain and sprocket drive. It was light for its time and on either petrol or paraffin could plough six acres in nine hours.

Two Ivels took part in the first official tests of the Royal Agricultural Society of England in 1910, along with three steamers and two Saunderson Universals (page 33). Of about one thousand Ivel tractors made, four are known to survive (one in the South Kensington Science Museum, one at Hunday National Farm and Tractor Museum, one in Western Australia – I know of this only because I met the owner at the Kendal Steam rally in 1981, quite by chance, so there could be others! – and one in Zimbabwe). The wonder is not that there are so few, but that there are any, for there must have been many middle years when these now valuable relics were just so many piles of rusting junk. Fortunately, space on farms was often ample and there was little reason to bother to clear out the corners of a stackyard or a distant barn. Admiration and praise is due to those who have the patience and knowledge to restore machines over seventy years old to as-new condition.

It is interesting to note that when the English Agricultural Society was founded in 1838 (the Royal Charter was granted soon after), 'men of science' were encouraged to give attention to 'the importance of agricultural implements'. The *Journal of the Royal Agricultural Society of England* noted that 'the farmer whose life is secluded has little opportunity to see . . . new machines', and this recognition caused the comprehensive trials organised by the Society.

Engine: Two cylinders horizontally opposed
22hp at 850r/min
Gears: One forward, one reverse
Dimensions: 9ft 9in (2.97m) long,
5ft 5in (1.65m) wide
Turning circle: 12ft (3.66m) diameter
Weight: 1 ton 8cwt (1.42 tonnes)
Price in 1903: £300 (US $1,461)

The Ivel tractor could be furnished with an engine cover, but ▶ it was an unlovely metal box, not often used, judging by contemporary photographs of the Ivel at work. Anyway, I could not resist the challenge to tackle the feast of chains, gear-teeth and concentric wheels in the engine of this precious antique. As the weight of the machine is concentrated at the rear, it seemed necessary to balance this with some bulk at the left of the background setting, this was provided by a limestone farm building based upon one seen at the village of Arncliffe in Littondale – the original location used for Yorkshire Television's series Emmerdale Dale (see p. 42). This is typical Yorkshire Dales country, beyond Upper Wharfedale and Airedale, the rough moorland of heather and bracken criss-crossed by drystone walls and cropped by sheep, only the valley bottoms supporting arable farms. For this painting, the open filigree patterns of mature trees with rooks nesting in their tops were preferred to the lush masses of summer foliage. The austere lines seemed to express the nature of this harshly beautiful country better and also provide contrast to the tractor's bulky square tank and complicated mechanical details.

INTERNATIONAL HARVESTER MOGUL 8-16, TWIN 12-25 · 1916

These two models shared the same engine, but for the smaller 8–16 one cylinder was removed and a blank plate was put on the crankcase. The 8–16 was the first International Harvester model imported into the UK, in 1915, the same year in which it won the top award at the Panama Pacific International Exposition in San Francisco.

The International Harvester Company of Chicago was one of the giants in the history of tractor development, formed in 1902 by the amalgamation of a number of implement makers, including Deering and McCormick, whose names were used on some models (page 35). Before the appearance of the Fordson it was easily the largest tractor-maker, being responsible for

THE IVEL

AO·385

J.H.APPLEYARD
1978

an output of 3000 in 1912, then one-third of the total US tractor production. It continues to make a wide range of agricultural machinery in many plants throughout the world, which included the old Jowett car works at Idle in Yorkshire, from which came the famous, horizontally opposed two-cylinder engine, made for over forty years in substantially the original form. This made it one of the longest lasting internal-combustion-engine designs of all time. There is a tractor connection, as this engine was one of those used in the Bristol crawler, made between the mid-thirties and the fifties.

Mogul 8–16

Engine: Single cylinder, 8–16hp
Gears: One forward, one reverse
Dimensions: 12ft 6in (3.81m) long, 5ft 6in (1.68m) wide
Turning circle: 20ft (6.10m) diameter
Weight: 2 tons 10cwt (2.54 tonnes)
Price: £350 (US $1,708)

Twin 12–25

Engine: Two cylinders, horizontally opposed, 12–25hp
Gears: Two forward, one reverse
Dimensions: 14ft (4.26m) long, 7ft (2.13m) wide
Turning circle: 25ft (7.62m) diameter
Weight: 4 tons 10cwt (4.57 tonnes)
Price: £580 (US $2,830)

INTERNATIONAL HARVESTER JUNIOR · 1919

As on some early cars, notably the Renault, the radiator of the Junior was mounted behind the engine, mainly to avoid accidental damage. Along with its larger relation, the 10–20 Titan (page 21), the Overtime (page 29) and the Fordson F (page 37), the Junior made an important American contribution to the production of food for Britain in the Great War. A total output of well over 30,000 Junior tractors in five years was significant, although dwarfed by the figures of 100,000 per year by Ford in the early twenties. Small

Tractors are not among the most elegant of vehicles and some of the earliest are classic examples of 'committee design', and this International Harvester Twin is one of the ugliest subjects I have painted. Large areas of steel plate and predominantly sharp corners produce a shape which suggests anything but movement. The large tree in the field where the Friesians graze serves to soften the angles of the high cab, and the tall front stack merges into the farm buildings. However, the main effect is of stark contrast between the hard lines of the machine and the soft contours of the landscape, with a stream which runs down this Nidderdale valley between two mills concerned with the processing of sheepskins. The most interesting features of the tractor were provided by the detail of the flywheel and pulley mechanism and the contrast between the deep green paintwork and the spindly, crimson wheels.

▼

The green Mogul in landscape produces a situation of colour ▶ harmony, but introduces problems of providing sufficient contrast. This was my first tractor painting and I soon realised how valuable were the thin open shapes associated with the red wheel spokes. Fortunately too, the range of tones in the tractor's paintwork is wide, from near-white highlights to deep-cast shadows. Plain areas of green are relieved by gold lettering giving the model number and details of the Grand Prize earned by this tractor.

The landscape is near Kilburn, the Yorkshire village where Robert Thompson, the 'Mouse Man', practised his craft of woodworking and carving. Behind rises the cliff at the head of Sutton Bank, one of the county's well known formidable hills at the southern end of the Hambleton Hills. Not shown is the famous landmark of the White Horse carved into the hillside and tended at intervals by the villagers of Kilburn.

J.H. APPLEYARD
1977

J.H. APPLEYARD
1977

This is one of my favourite views in North Yorkshire, across the valley of the Leven (a tributary of the River Tees) near Crathorne Hall, towards the Cleveland Hills. The post-and-rail fence replaces an actual hedge which is smothered in summer with convolvulus, but this would have confused the detail of the Junior and, despite its beauty, had to be simplified.

The tractor has a simple, pyramidal shape containing a series of details of varied form – the highly finished shoe-shaped bonnet contrasts with the square radiator block and header-tank; the horizontal cylinder of the fuel-tank sets off the milkchurn form of the reservoir, and the steering wheel and spaces around the spring seat give a lightness to the rear end. Grey and dull red look well in the greens of landscape.

Some tractors which are simple in outline require more detail in their setting to compose a satisfactory picture, but the Titan conceals nothing and is portrayed against the hills of Wensleydale and a north-country gin-gan; the original of this can be found at the Beamish Open-air Museum in County Durham. This pyramid-roofed building (sometimes conical) housed an engine (of which 'gin' is a northern dialect version, also 'cotton-gin') powered by from one to four horses which were tethered overhead to a system of levers and gears for powering machinery for purposes such as threshing. One of the horizontal bars to which the horses were fastened can be seen between the stone pillars.

The primary interest in the composition of this tractor lies in the contrast between solid and void in the upright ellipses of the wheels and pulley mechanisms and the horizontal and vertical cylinders at the front of the tractor. Colour is very low key, but the red on the wheels and the logo of the International Harvester Company introduces some variety.

farmers found the 30cwt (1.52 tonnes) Junior an attractive proposition at £300, especially after demonstrations of its capabilities at the South Carlton trials in 1919.

A typical trial programme included seven classes: internal-combustion-powered direct traction of various horsepowers; steam direct traction; steam or internal-combustion cable ploughing and self-propelled ploughs (see page 23). Each class was set to work both light and heavy soils, had to stop and start uphill and downhill and there was a test for belt pulleys in actual driving. Turning circles and weight were taken into account in assessing the tractor's performance.

The International Harvester Junior was the first model to be designated 'International' and the first tractor to have power take-off by shaft as an option. An import figure of approximately 2,500 bears witness to its popularity on arable farms in Britain.

Engine: Four cylinder vertical, 24hp at 900r/min
Gears: Three forward, one reverse
Dimensions: 11ft (3.35m) long, 5ft (1.52m) wide
Turning circle: Unobtainable
Weight: 1 ton 10cwt (1.52 tonnes)
Price: £300 (US $1,305)

J.H. APPLEYARD
1077

INTERNATIONAL HARVESTER TITAN 10-20 · 1918

The 10–20 Titan was one of the last successful primitive tractors where a stationary-type engine was mounted on a girder frame. With its forward mounted water tank it looked much like a steam engine without a chimney. It was familiar to British farmers as over 3000 had been shipped from America to help with wartime food production. Both facts had some bearing on the reduction in price in 1919 by $225 to $1,000 (£285), but its price did not remain competitive with the Fordson F (page 37) for long.

International made it clear that the reduction was not due to incomplete equipment and their advertising suggests that many other companies were offering machines that were bare prime movers and that often the farmer was left with no idea how to start or work his new machine unless he was prepared to pay for a course of instruction.

Along with other models from International Harvester, the 10–20 Titan was at the South Carlton Trials, organised by the Society of Motor Manufacturers and Traders and the National Farmers' Union in 1919, where an attempt was made to demonstrate and test a wide range of available machines. By 1919 International could claim 88 years in the farm machine business and an involvement with tractors since 1905. Their expertise has stood the test of time, for the company which made this '3 plow kerosene tractor' of 1918 is still making farm machinery.

Engine: Two cylinder, 20hp at 500r/min
Gears: Two forward, one reverse
Dimensions: 12ft (3.66m) long,
      5ft (1.52m) wide
Turning circle: Unobtainable
Weight: 2 tons 15cwt (2.80 tonnes)
Price in 1922: £410 (US $1,543)

CRAWLEY MOTOR PLOUGH · 1918

This extraordinarily cumbersome-looking machine was operated by one man seated behind, in a similar way to the modern garden rotovator. Considering its size and weight, it was probably no more difficult to handle than a team of horses. In fact, it is a simple and obvious result of the idea of replacing horses by an engine. It was exactly as a mechanical horse that the tractor was regarded for the first twenty-five years of its existence, some even being operated by reins to make the driver feel at home.

Like the International Harvester Junior (page 20), it has its radiator behind the engine to avoid damage and the engine cover is remarkably similar in shape to the shoe-shaped bonnet of the veteran Renault car. One advantage of this 'Agrimotor' was that the work was in front of the operator so that he could easily see what he was doing. Using a draught plough required a second man riding on the plough or the tractor driver had constantly to look back to check the operation.

The first Crawleys were built by the famous agricultural machine company, Garrett of Leiston, before the Crawley Bros' own factory was set up at Saffron Walden. The Crawley won its class at the 1919 tractor trials, but motor ploughs were clumsy, and lacking versatility, could not compete with the many rapidly developing tractors on the postwar market.

A frame with a wheel could be attached to convert the Crawley into a tractor and in this form it could be used for hauling a reaper and binder.

Engine: Four cylinder vertical (Buda)
        24hp at 1000r/min
Gears: Two forward, one reverse
Dimensions: 17ft 6in (5.33m) long,
      5ft (1.52m) wide
Turning circle: 24ft (7.32m) diameter
Weight: 2 ton 6.5cwt (2.36 tonnes)
Price in 1919: £545 (US $2,370)

This motor plough was unrestored when seen so I had to ▶ *visualise how it would look when in new working order, including the colours, working out shades as exactly as possible. In order to make the most of the machine's form, which is not especially elegant, I chose to show it in bright sunlight from the right, giving strong contrasts between the shadowed sides and sunlit paint and reflections and also to make an attractive pattern of light and shade on the rims and tubular spokes. The visual responses are quickened by the play of light.*

Above the surrounding cottages and the farm building with its outside stairs rise the fells of Upper Wharfedale crossed by the characteristic limestone walls and showing outcrops of this white rock. The houses are part of the small village of Buckden, near Kettlewell, situated in the meagre arable strip at the bottom of the narrow valley.

J.H.APPLEYARD
3/79

J.H. APPLEYARD
1978

◀This is a very cold painting – blue tractor with blue-grey wheels, shadows on grey concrete and bare trees in sharp, winter light. Only the sandstone wall and the implied warmth of the sunlight relieve this feeling of chill in the air. Perhaps the Weeks is rather out of its element and could have been better shown against hop-gardens and oast-houses, but this is outside my recent direct experience and I chose to paint it where I saw it. The relatively simple details of the tractor, especially with the bonnet side-plate in position, persuaded me to use a busy setting where the verticals of tall, slender trees rising out of the valley catch a strong, low sunlight from the left. I always look for this pattern element in deciding what background is most suitable, for example, lacy trees against a plain, solid machine; square red tractor against the rounded shapes of green, summer trees; horizontal tractor, firm uprights of winter trees, and so on. This conscious selection and placing must, however, avoid appearing to be too contrived.

The Wallis Cub Junior is particularly interesting as a painting ▶ subject in this series because of its unusual colour scheme of dark blue and buttercup yellow. Also unusual, but not unique, is its tricycle wheel formation and the bear-cub mascot above the front wheel. This mascot, clearly referring to the model's name, indicates the degree to which the front wheel is turned – a simple and ingenious safety device (shared with the Gyrotiller page 38), fitted because the driver's position gives no view of the front wheel.

This tractor has attractive clean lines depending upon its unitary construction, with its dark blue bonnet sitting low between the yellow wheels of simple, flat-spoked type. The range of colour on the wheels varies from pale, cold lemon where light from the sky reflects, to a warmer, darker tone under the upper rims. These darker tones of yellow always present an interesting technical challenge in paint mixing, as dark yellows cease to be yellow.

The setting is once again farmland beside the River Tees, where a makeshift fence leads into a more formal hedge with trees ready to spring into leaf, bordering a field dusted with green, shooting cereal. Its perspective leads back into the mysterious blue-grey distance, echoed in colour by the reflecting bonnet top, all seen in the bright, but diffused, light of a spring day with thin cloud and patches of blue sky.

WEEKS-DUNGEY
NEW SIMPLEX MODEL · 1922

With its wide variety of climate and terrain, the USA provided conditions for many different crops. As a result the American tractor industry was required to produce many machines for specialist tasks; the row-crop tractor with high ground clearance and differing wheel arrangements is perhaps the most familiar.

In Britain, the general-purpose tractor could cope with most farming jobs, so the Weeks-Dungey

machine was a rare early exception. It was a light and narrow tractor designed specially for use in the Kentish hop-gardens by a Mr Dungey and put together mainly from mowing-machine and car bits. His design, in 1914, resulted from a vain search for a tractor suitable for his requirements in hop cultivation. The local blacksmith was persuaded to build the machine using a proprietary American engine (Buda/Waukesha) and its success was notable enough for Mr Dungey's neighbours to ask for copies to use on their own farms. Over a period of about ten years of developing design just over two hundred were built by W. Weeks & Son,

Perseverance Iron Works, Maidstone, who were also agents for the Angus-Sanderson car. Two examples of early models are known to exist, one at Wye College's Agricultural Museum at Brook and one privately owned, both appropriately in Kent; the latter example dug up in a Kentish orchard.

Several interesting features were introduced, including a differential lock on the drive wheels and a foot accelerator. A set of spare road wheels with rubber tyres cost £70. Tractor production ceased in 1925, but the Weeks company is still in existence producing spraying machinery.

Engine: Four cylinder vertical
       25hp at 900r/min
Gears: Three forward, one reverse
Dimensions: 8ft 6in (2.59m) long,
       4ft (1.22m) wide
Turning circle: 25ft 9in (7.85m) diameter
Weight: 1 ton 15cwt (1.78 tonnes)
Price: £485 (US $1,826)

WALLIS CUB JUNIOR · 1916

Both the Wallis Cub and the Cub Junior were frameless tractors, appearing in 1913 and 1915 respectively. There was a rolled-steel sump for the engine and gearbox and this boiler-plate design remained with Massey-Harris until the Second World war. In terms of those important factors price and weight, the effect was all beneficial, but farmers remember that access to the engine and gearbox was much easier on the older type of tractor where the motor was mounted on an angle-iron frame. The disappearance of this girder chassis transformed the tractor and unit construction, of which the 1917 Fordson (page 37) was a classic example, set a style which is largely unchanged and predates most chassis-less cars by about forty years.

The rear wheel gears of the Cub Junior were lubricated by oil and carbon from the exhaust pipes and the pivoting front wheel gave a good turning circle. Claims that it could turn in its own length seem unlikely, especially if spade lugs were fitted to the rear wheels, although without stops the front wheel could be turned to 90° from the longitudinal axis.

The J. I. Case Plow Company was run by a branch of the family which made Case Tractors at the J. I. Case Threshing Machine Company. 'Case Plows' made Wallis tractors and was eventually taken over by Massey-Harris, relinquishing all rights to the use of the name Case.

Engine: Four cylinder vertical
       13–25hp at 850r/min
Gears: One forward, one reverse
Dimensions: 11ft 9in (3.58m) long,
       5ft 6in (1.67m) wide
Turning circle: Unobtainable
Weight: 1 ton 8cwt (1.42 tonnes)
Price in 1916: £420 (US $2,049)

WALSH & CLARK VICTORIA OIL PLOUGHING ENGINE · 1918

Many of the first horseless carriages and the early railway coach are classic examples of the tendency to base the layout and style of new machines upon the appearance of the apparatus that they were intended to replace. Although Walsh & Clark of Guiseley in Yorkshire had earlier made oil-engined tractors, in 1915 their oil-ploughing engine was redesigned and became virtually a steam engine with an oil engine mounted on top. It was used in just the same capacity as the cable-ploughing steamer.

Cable ploughing started with a patent of 1811 and

This traction-engine-like machine was built to haul a plough, but by cable from the horizontal drum seen under the 'boiler' and not by direct traction. It is shown at the rear entrance to one of the Hunday display halls and is of special interest to me as it was built at Guiseley, some 10 miles from Leeds and one of my childhood haunts; still the home of Harry Ramsden's, perhaps the world's most famous fish and chip restaurant.

As the surrounding trees are silhouetted darkly against the sky, the only colour contrast comes from the red of the road wheels and flywheel, but the range of tones and colours in the foliage are the most interesting feature of the background to me. Colours stretch from a near-brown russet green to a dark prussian blue; tones from the cool blue-green of leaves reflecting the light to warm blacks in the tree structures. Because I paint so much landscape, I consider that green is the most versatile of all colours, edging here into metallic blue-greys where the engine's paint picks up colour from the sky. Oddly enough, I find that the varied shapes of sky, revealed between the leaves, work best in paint if they are put in after the dark areas of leaves have been painted, when one might expect that the leaves are painted over the sky.

J.H.APPLEYARD

was developed by John Fowler's design of 1854, which was the first of a range of steam-ploughing engines used world-wide for over sixty-five years, although the method was little used on the massive farms of the USA. Both Fowler and McLaren (whose works were side by side in Leeds) demonstrated cable-ploughing at the Aisthorpe trials in 1920, although by then the power was no longer steam, but petrol and/or paraffin.

An engine was stationed at each end of the field to be ploughed. From the engines' winding drums 900yd (820m) of ¾in (19mm) steel cable was connected to a 5/6 furrow balance plough or a reversible cultivator. As engine A pulled the working plough across the field, engine B was stoked with coal and moved a small distance along the headland to line up for the next pull. A very small area of unploughed headland remained, as the raised ploughshares finished right over the working engine. Then the operations were reversed. The other set of ploughshares was lowered, engine B took up the pull and engine A refuelled and moved. And so on alternately. Two hours were needed to raise steam and a 4.30am start was usual for the contractors from whom the machinery was usually hired, being too expensive for the average farm (£1,000 or more, US $4,880).

The Victoria engine was a clear case of the retention of obsolete layout and style to influence the conservative farming community and help sales. Some two-stroke models were offered, but most, including the example shown, used the Otto four-stroke cycle.

Engine: Two cylinder horizontally opposed, side valve
45hp at 500r/min
Gears: Two forward, one reverse; one cable speed
Dimensions: 14ft 6in (4.42m) long,
6ft (1.83m) wide
Weight: 6 tons (6.10 tonnes) with fuel and water
Price: £1,000 (US $4,880)

WATERLOO BOY MODEL N (OVERTIME) · *1916*

Any tractor which gained the support of that uncompromising gentleman, Harry Ferguson, must have shown commercial promise and some technical ingenuity. So the fact that he opened the first UK agency for the Waterloo Boy soon after the Great War (long before he designed his own famous tractor) indicates a machine of above average interest.

In an Iowa town named after his family, John Froehlich built two self-propelled, reversing tractors in 1892 and sold them both. Unfortunately, both were returned by dissatisfied customers. The Waterloo Gasoline Engine Company was founded by Froehlich and introduced new models in 1896 and 1897 but sold only one of each, relying upon stationary engines for their income. It was not until 1912 that the first Waterloo Boy was introduced and then in 1915 the Model N appeared. By 1918 the company had become the John Deere Tractor Company and green and yellow paint was adopted.

'A Giant in Power A Miser in Fuel' as a slogan strikes a familiar note to the reader of the late-twentieth century, but was in fact written in 1916 to emphasise the claimed efficiency of the Waterloo Boy 'motor'. This was due mainly to complete combustion of fuel and power was transmitted to the rims of the rear wheels where the drive gear was shielded from dirt.

In 1918 the government offered a prize for ploughing against the clock which was won by a team using a Model N. Most significantly the team consisted entirely of land-girls, still showing their ability to cope with any job on the land, as they had been doing for four years of war (officially as the Women's Land Army since 1917).

The descendants of this pioneering company still wear yellow and green in farm fields throughout the world (page 51).

Engine: Two cylinder horizontal
12–25hp at 750r/min
Gears: Two forward, one reverse
Dimensions: 13ft (3.96m) long,
6ft (1.83m) wide
Turning circle: 22ft (6.70m) diameter
Weight: 2 tons 12cwt (2.64 tonnes)
Price: £325 (US $1,586)

In this painting, unusually I have gone for a clearly picturesque ▶ situation – water and beautiful trees for their own sake. I saw the setting by the lake at Westside, near the Hunday Museum in Northumberland, on a glorious summer's day and knew that the landscape would stand by itself with its many changes of tone, colour and texture produced by the grouping of conifers, silver birch, water-side willows and dense woodland of beech and chestnut. The juxtaposition of trees and water reminds me of Monet, and although I am not consciously influenced by other painters in the way I paint, I admit to great debts in the way I see to masters such as Van Gogh, Bonnard, Paul Nash, Andrew Wyeth and Stanley Spencer's intense examination of natural detail. Opinions will differ as to the wisdom of showing the primitive Waterloo Boy (Overtime) tractor in this setting. The tractor is a perfect example of the result of placing an oil engine upon a simple girder chassis, with some consideration of practicality (as in the sideways-mounted radiator to avoid sucking in dirt), but little, if any, thought given to appearance. I felt that the association of crudity and natural beauty was mutually enhancing. Oddly, if I saw such a machine left in this position in real life, I would feel it was out of place. However, see the Introduction for further thoughts on the relationship between art and beauty; better still read The Meaning of Art *by Herbert Read, a fellow Yorkshireman.*

CLAYTON-SHUTTLEWORTH · 1918-27

If you are a keen gardener you will recognise the warnings that professionals issue every spring about the problems caused by working too promptly on wet soil while trying to get crops like peas going early. Compression of the soil surface into a hard pan outweighs any advantage of early sowing, but the use of a plank to walk on helps to spread the gardener's weight over a larger area, on the same principle as snow-shoes, and thus decreases the pressure per square unit on the soil. The chain-rail, tracked or crawler tractor applies the same idea to the agricultural machine, not by any decrease in the machine's actual weight, but by spreading it over wide, endless tracks which offer a much greater area than the tangential contact from conventional wheels, whether steel or rubber-tyred.

The Clayton Chain Rail Farm tractor, as the company of Clayton and Shuttleworth of Lincoln, England called it, put down a surface on each track of 5ft 10.5in (1.79m) long × 1ft 2in (0.35m) wide, giving a total area of 1,974in² (12,532cm²), which for a tractor of 56cwt (2,845kg) exerts a pressure of just over 3lb per in² (0.21kg per cm²). This compares favourably with a pressure of approximately 2lb per in² (0.14kg per cm²) applied by a 12 stone (76kg) man standing firmly upon two size ten feet. The Clayton was steered by clutches on the countershaft and is recorded by the makers in their publication No 341 as having ploughed 4,000yd² (3,344m²) of barley stubble to a depth of 5.5in (14cm) using a three-share Moline plough in one hour on 3.36gal (15.27 litres) of paraffin.

The tracklayer idea is over two hundred years old, a patent having been granted in 1770 to a Mr R. L. Edgeworth; Boydell's 'Endless Railway' dates from 1846 and the first agricultural machine to use the idea was built by R. Bark of Birmingham, England in 1855. The crawler tractor was never very popular on European farms, but there was a surge of interest in the fifties with a number of firms producing tracked machines and conversions (see page 47 and illustration on page 61). The current importance of the crawler lies in its development into large, complicated machines for the civil engineering industry.

Engine: Four cylinder
　　　　35hp at 1000r/min
Gears: Two forward, one reverse
Dimensions: 11ft (3.35m) long,
　　　　5ft 4in (1.63m) wide
Turning circle: Clutch/brake steering
Weight: 2 tons 16cwt (2.84 tonnes)
Price in 1919/20: £385 (US $1,500)

GLASGOW · 1918

In its time the Glasgow tractor was as technically unconventional as the organisation of the company which produced it. This model by the D.L. Motor Manufacturing Company of Motherwell has a Waukesha engine driving all three wheels, the front pair slipping on ratchets in place of the usual differential gearing. Deeply dished pressed steel wheels and quite advanced, solid, unit construction contributed to a heavy, durable appearance, but the tractor weighed under 2 tons (2 tonnes) and was particularly suited for work on wet land.

Two other companies, J. Wallace & Sons and Carmuirs Iron Foundry, combined with D.L. in 1919 on an ex-government site at Cardonald to mass-produce the Glasgow tractor. The venture was to be organised as a profit-sharing scheme and some 2,500 workers were to turn out 5,000 tractors a year. Only 200 ever left the factory.

It seems that the marketing company concerned, The British Motor Trading Corporation, over-extended itself by the advance purchase of the first five years' production. The corporation's connection with the 11.9hp Bean car did not help, for this ambitious attempt to mass-produce a popular car never achieved one-tenth of the proposed 50,000 cars per annum. Ford and Morris proved that the buyers were waiting, but there is clearly more to success than identifying potential customers.

Engine: Four cylinder vertical
　　　　27hp at 1150r/min
Gears: Two forward, one reverse
Dimensions: 11ft 4in (3.45m) long,
　　　　5ft (1.52m) wide
Turning circle: 36ft (10.97m) diameter
Weight: 1 ton 17.5cwt (1.90 tonnes)
Price in 1920: £550 (US $1,920)

Crawler tractors are seldom used on the farms of Durham and North Yorkshire, the areas I know best, so to set the Clayton tractor in typical surroundings I found it necessary to use reference material depicting the flat lands of Lincolnshire or East Anglia. Distances seen are not great and the features of this level countryside overlap and are compared with the vertical quality of the picturesque windmill, its open sails and wind-vane contrasting with the solid, dark mill-tower. The pearly colours of an evening sky are reflected in the straight perspective of the drainage ditch, which serves to push the dominating mill back into the middle distance, only the declining sun and its warm reflected light complementing the deep greens of the foreground and the tractor. The latter is a pleasantly balanced, business-like machine with mechanical details and controls contrasting with areas of steel plate. An eccentric feature of the tractor shown by the imprints in the wet track are the horseshoe-shaped ridges on the plates of the crawler track – an odd hankering after the old days of horse traction.

J.HAPPLEYARD
12/82

SAUNDERSON UNIVERSAL MODEL G · 1916

A novel use was found by the Jockey Club at Newmarket for a Universal when it was employed in 1911 for grass cutting, harrowing and hauling a 10ft (3m) roller weighing 3 tons (3 tonnes).

The Elstow works of Saunderson and Gifkins produced a wide range of machinery in keeping with their trade name 'Universal'. Besides tractors for hauling and driving machinery, they offered stationary and portable engines, both oil and petrol, and others for marine use. Their weed-cutting machines were in use as far afield as Spain and Egypt (for sudd cutting) as well as in the British colonies. Tractors ranged from a single-cylinder, air-cooled model to four-cylinder, water-cooled models rated at 50hp, with pumps and wind-motors completing an impressive list of products.

In 1910, Saundersons competed with two Ivels (page 17) in the first trials organised by the Royal Agricultural Society of England at Baldock, although at this time the judges preferred steam-powered entrants. However, Saunderson collected a long list of awards, including an RASE silver medal in 1906, and claimed never to have been beaten in any competition.

This Nidderdale composition consists of two overlapping ▶ triangles, one upright, one inverted, the stone wall and the strutted fence-post being clear pointers to these main divisions. Other minor triangles are superimposed, including those which move into space, such as the skyline, the farm road and the clump of roadside grass behind the rear wheel. Dark groups of trees serve to frame the tractor and to lead the mind out into the world beyond the picture.

The busy, exposed structure of the tractor emphasises the verticals in the radiator and the exhaust stack, contrasting with the many circles, mainly red, including the wheels, pulley and flywheel. A broken shadow from outside the picture format is used to break up an otherwise empty foreground and contributes to the repeated triangular theme.

The details of this setting are very familiar to me, consisting mainly of parts of the view from the front window of the bungalow where my mother-in-law used to live.

◀ The unorthodox Glasgow is painted against a backdrop of Swaledale scenery, James Herriot country, above the village of Reeth, a section of one of the steep-sided valleys carved by the eastward flowing rivers of North Yorkshire. Standing on rough pasture, capable of supporting only the hardy moorland breeds of sheep to one of which the dale gives its name, the tractor is at the limit of its usefulness, for the ploughing of the moors has long been a controversial issue.

In the painting, space is suggested by the progressive blueing of the hills with distance and the glimpse of the horizon between the tractor's large header-tank, which forms the curved bonnet top, and the engine. It might seem that the conventional layout of bulky front engine and contrasting rear end of steering wheel and seat, combined with the common green and red colour scheme, would be boring, but here the extraordinary cast wheels proved to be diverting. The challenge lies in the everyday problem to the painter of creating the illusion of depth in these dished discs on a purely two-dimensional surface. A full range of crimson reds, from near black to cool mauve-pink, is needed to produce the impression of protruding centres, recessed inner sides and the almost spherical rear wheel. They remind me of giant-sized versions of the wheels fitted to the old Triang children's tricycle – do you remember the blue and red toy with the pedals each side of the front wheel?

J.H. APPLEYARD
9/78

J.H. APPLEYARD 1/79

The mid-range Model G, rated at 25hp, lasted at least fifteen years from 1916 and demonstrated its ploughing ability at a fuel cost of 4s 4¼d (US $0.77) per hour in the 1920 Lincoln trials. Strong, adjustable ploughs with assisted lift were also manufactured by Saundersons, who pointed out that ploughing constitutes one-third of the annual value of a tractor to the farmer. In 1919 the company traded under the name of the Saunderson Tractor and Implement Company and Universals also were marketed from Saunderson and Mills.

In 1925 the company succumbed to the depression and were taken over by the stationary-engine firm of Crossley Bros and thereafter the tractors were generally called Crossleys, production ceasing completely in 1932.

A 'useful' book of working instructions with a special section on 'Methods of using implements' was offered for 1s, plus 2d postage (US $0.17 plus $0.03 postage).

Engine: Two cylinder vertical
    25hp at 750r/min
Gears: Three forward, one reverse
Dimensions: 12ft (3.66m) long,
    5ft 6in (1.68m) wide
Turning circle: 36ft (10.97m) diameter
Weight: 2 tons 12cwt (2.64 tonnes)
Price: £510 (US $2,488)

McCormick-Deering International Harvester 10-20 · 1927

Visibility of the land ahead was frequently a problem with early tractor designs and various ideas were tried to improve the driver's view. He could be placed in a forward control position as on Garrett's 'Suffolk Punch' steamer, but this must have given a very poor view of towed implements and the land behind, not to mention the isolated operator of a riding plough. The engine could be offset, but perhaps the simplest solution was to offset the driving position as on this 10–20 McCormick-Deering, also sold as an International. For the later Farmall range International Harvester adopted the name 'Cultivision' for the idea. Conventional unit-construction, built-in power take-off and overhead-valve engine made this a popular and versatile mid-range tractor as shown by production figures of almost 216,000.

The 10–20 was offered as an industrial tractor with over 70 companies making accessories and equipment. The machine could be adapted for various towing jobs, road rolling, as a crane and for many operations such as sawing, pumping and stone-crushing, where it provided static power only. The Highway Trailer Company of Edgerton, Wisconsin, claimed that the use of their 'garbage collecting units' in conjunction with 10–20 tractors saved $2,000,000 per year in Chicago. The Street Cleaning Department of New York City employed 10–20s fitted with cabs for snow ploughing duties, the automatic 'plow' being controlled from the driver's seat.

The Austin Motor Grader (shown on a larger 15–30 page 2) and other machines of this kind were the most spectacular adaptations, with the original tractor almost unidentifiable as part of a complicated machine up to 7.6m in length. The leaning front wheels counteracted the tendency of the machine to move sideways

when the scraper blade was lowered into contact with the road surface.

Engine: Four cylinder vertical
    17–23hp at 1000r/min
Gears: Three forward, one reverse
Dimensions: 10ft (3.05m) long, 5ft (1.52m) wide
Turning circle: 24ft (7.32m) diameter
Weight: 2 tons (2 tonnes) approx.
Price: £220 (US $1,069)

Most rivers offer some variety of landscape on their way from ▶ *source to confluence or the sea; the northern rivers which rise on the slopes of the Pennines frequently flow into heavily industrialised estuaries. The Tees is a perfect example, starting as a clear stream on the wild moorlands of north-west Durham, flowing over Cauldron Snout and cascading from the rocks of High Force into a gorge which flattens into miles of rolling green farming country, until some 15 miles (24km) from the sea it is completely urbanised and industrialised. By now, the water is literally poisonous with effluent and sewage (we are told it is being improved to the point where salmon could once again be seen) and the banks are littered with all the repellent aspects of industry, iron and steel, chemicals, oil and gas, rendered worse by some of the depressing dereliction of economic recession. The latter contrasts vividly with this buttercup-dotted field, by the Tees, near a common called The Desmesnes at Barnard Castle, a market town boasting a medieval fortress, an eighteenth-century market cross and town hall and the Bowes Museum – a mock French chateau associated with the Queen Mother's family, housing quirky Victorian curios and masterpieces of European painting.*

This scarlet McCormick-Deering is placed in a patch of shadow producing a classic contrast of complementary colours, its simple shape clarified by the dark hedge and distant group of trees. Highlights on the tractor's paintwork are the effect of reflected light, giving a range of subtle purple hues on the bonnet top, pulley and mudguards and dull orange-brown on the wheels produced by the mixture of red and reflected green. Contrasts of shape come from the square tractor and the soft, rounded forms of the summer trees.

McCORMICK-DEERING

JH APPLEYARD
8/78

FORDSON MODEL F · 1917

This tractor revolutionised the tractor industry. It was first produced in 1917 by Henry Ford at Dearborn and was the result of many experiments by the Ford design engineers to produce a tractor. It was the first unit-constructed, frameless tractor and was built using the same techniques that Ford employed in the mass-production of the Model T car. The weight of just over 1 ton and the 20hp engine made it more suitable than most for a wide range of work on the farm; the average 20hp tractor during the Great War would have weighed 3 or 4 tons. With the manufacturing techniques used it was possible to undercut the price of most rivals and the Model F was the direct cause of many manufacturers going bankrupt or stopping the production of tractors.

The first 7,000 off the production line in 1917/18 were exported to Britain, which was facing a very serious food shortage due to the submarine blockade by Germany and the loss of many thousands of horses and men, formerly producing food, now fighting and dying in France. When the tractors reached this country they were distributed by the Ministry of Munitions and many were operated by land-girls who worked in shifts to keep the tractors ploughing twenty-four hours a day.

This tractor was readily accepted because of its low price and general reliability, but like the Model T Ford car, although the model never changed, nearly every component was altered during the course of its eleven years of production which includes three years of manufacture in Northern Ireland from 1919 to 1922. Over the eleven years nearly three-quarters of a million were produced, making it one of the best selling tractors of all time.

Prices quoted in various trials reports are £280 (US $1,218) in 1919, £205 (ex works) (US $772) in 1921 and £143 (US $686) in the mid-twenties (ex works, Trafford Park). The Fordson Model N sold for £155 (US $753) in 1929 and had risen to £178 10s 0d (US $720) in 1945 (see page 48).

Engine: Four cylinder
11–20hp at 1100r/min
Gears: Three forward, one reverse
Dimensions: 8ft 6in (2.59m) long,
5ft 2in (1.57m) wide
Turning circle: 22ft 6in (6.86m) diameter
Weight: 1 ton 4cwt (1.22 tonnes) approx
Price in 1920: £260 (US $908)

FOWLER GYROTILLER MODEL 170 · 1935

Fowlers of Leeds had been in steam since the 1860s, but when steam power finally declined in the twenties the firm needed a product to fill the gap before the introduction of their crawler tractors. The first example of the Gyrotiller appeared in 1927 (the company were also making motor ploughs), developed from the ideas of an American called Storey, who had unsuccessfully peddled the design around US manufacturers, including Caterpillar. Sales of the massive, complicated machine were slow, but had reached three figures by 1935. Being both large and expensive the machines, like the old steam ploughing set, were operated mainly by agricultural contractors. They were ideal for deep cultivation and the clearance of scrub, but were often set too deep, damaging the sub-soil, although they were successfully developed for use on the sugar-cane plantations in the West Indies, where their use was to pulverise and aerate the soil which was claimed to reduce germination periods.

Two tiller rings, each fitted with alternating cutting shares, rotated on a vertical axis (which could be tilted to lift the cutters), operating like the beaters of an egg-whisk or a food-mixer. Degrees of pulverisation of the soil could be varied by altering the ratio between track and tiller speeds and the number of cutters on the rings.

Gyrotillers came in four sizes with engines of 30, 40, 80 and 170hp, usually MAN or Fowler-Sanders diesels, which replaced the original thirsty Ricardo petrol engine. The three-cylinder 30, four-cylinder 40 and six-cylinder 80 had non-steering front wheels, while the largest model carried a windvane-like indicator to show the driver the position of the steerable, flanged front wheel (see also page 24).

A booklet, compiled in about December 1935, quotes many testimonials to the 'enormous advantages to be gained by "Gyrotillage" ', and special emphasis is placed upon the ability of 'Gyrotilling' to break up the ploughpan and eradicate deep-rooted weeds.

Production ended in 1938/9.

Engine: MAN diesel, six cylinders
170hp at 400r/min
Gears: Two forward, one reverse
Dimensions: 26ft 3in (8.00m) long (without ridgers),
11ft 1in (3.38m) wide
Turning circle: Unobtainable
Weight: 30 tons (30.5 tonnes)
Price: Model 80, £4,000 (US $19,440) in 1931,
Model 170, £6,000 (US $29,400)

The furrows of ploughed land are a Godsend in helping to create ▶ *the feeling of distance and space which is one of the main pre-occupations of the western realist painter. Allied to the lines of the verges and hedge and the diminishing sizes of the trees, the ploughed section of the fields makes a strong movement from front right to left rear, broken by the plough's control handles to form a pyramid. The general lack of colour in the grey Fordson places emphasis upon the tractor's orange wheels and the pastel blue framework of the Ransomes plough.*

This view is so close to my home that I can almost see it as I write. A late-season activity on the farm is shown, but not yet late enough for the trees to be wearing autumn colours; the hedge has been trimmed and laid and the already fading rough vegetation around the tractor indicates that this is a field not recently cultivated. Once again the soil is light and dry enough not to have clung to the tractor's ridged and transversely cleated wheels! (See pages 52 and 64.)

JHAPPLEYARD - 12/82

◄ *Fowler Gyrotiller 170*

This massive, complex machine standing over 13ft (4.0m) high has sufficient detail to sustain a composition without compensating features in the setting. So it is shown as seen in action, surrounded by clods of freshly turned earth, at a ploughing demonstration in the grounds of the Hunday Museum in Northumberland. The inclusion of the operator's figure sets the scale by which the size of the machine can be appreciated.

Every inch of the machine bristles with interesting features to paint, from the roller-like front wheel, with its pennant, coloured white on one side and red on the other, to indicate which way the wheel is turned, and the cylindrical tank, to the rear platform supporting the plough tines, here shown raised, as the machine turns, to reveal their shape. Red paint alternates with pristine green areas, raised yellow lettering, chromium, polished steel, brass and copper pipes and wooden mouldings. Besides the imposing diesel engine, two parts of special interest are the undercab details of wheels, chains and levers and the red wheels and green structure of the track mechanism, partly hidden by a heavy H-section girder.

Repetition of a subject can often become boring as it tends to remove the element of discovery in the process of painting, but despite having made three paintings of the Gyrotiller, I rate this one of the most absorbing tractor subjects I have tackled. I hope I have captured some of the thrill of seeing this magnificent white elephant in action.

Most of the interest in this painting lies in the rotary hoe so predictably I chose a rear view, perhaps undervaluing the tractor element a little. However, the colour scheme and the close-coupled proportions of the machine make it clear that this is an integrated unit.

The strong sunshine from over the observer's right shoulder creates contrast between light grey highlights and black shadows, particularly on the hoe blades, which are here shown raised. It is perhaps necessary to remind the reader that the section of green field, receding under and behind the tractor, is actually, on the painting, a vertical area. The illusion of distance rests upon two related factors – the smoothing out of texture with distance coupled with the gradual decrease in tonal contrast. Of course, the fence, haystacks and buildings are also pushed back by the simple and obvious device of overlapping tractor parts, such as the exhaust stack and steering wheel, with background features.

HOWARD MODEL DH22 · 1928

I suspect that the average Britisher's picture of Australian farming would include a lean, sun-tanned, hard-drinking man herding millions of sheep and struggling against alternate drought and torrential rains, rabbits, kangaroos and flies. It is an impression of a life which is worlds away from the routine of an English arable farm and few of us would expect it to be an activity that would encourage not merely the use of farm machinery but the development of a new type of cultivator.

However, an Australian called Arthur C. Howard, known as Cliff, produced a prototype rotary hoe in New South Wales in 1920 and great claims were made for its performance compared with the drawn plough. Its 60hp Buda engine drove the tractor wheels and the rear, horizontal rotor (Howard's spelling in the DH22 instruction book). This consisted of a number of blades attached by flanges to a tubular spindle; the handbook added 'The thrust caused by the resistance of the ground against the blades assists in propelling the machine forward' and it claimed to cultivate 3.5 acres an hour. Sales began in 1922 of this first commercial rotary tiller from Austral Auto Cultivators of Northmead, New South Wales, but the most successful model was the DH22 which appeared in 1928. Two versions were offered; one for sugar-cane work,

J.H. APPLEYARD
1/80

adapted for working as much as 14in (0.35m) deep and the field/orchard model which worked 'the land up to 7in deep'. Bent or worn hoe blades seriously affected the implement's efficiency taking 'twice the power to drive' and included in the standard equipment was '1 bar for setting blades', kept in the cultivator's tubular spindle. On account of the dusty conditions in which a rotary hoe must work, air to the carburettor was passed through an oiled horse-hair filter which required weekly washing in kerosene.

The rotary hoe could be detached from the tractor and a completely mechanised system for cereal growing was planned, including seed drills and harvesters.

Instructions for starting the engine were conventional – retard the spark, pull out choke, turn on petrol, ignition off – then three or four pulls on the starting handle sucked in fuel. Choke in, ignition on, 'then give the starting handle a couple more pulls upwards! . . . when the engine is running the ignition is advanced. After a few minutes the fuel control tap can be switched across to kerosene.' The starting handle could sometimes be removed but on most tractors was fixed and needed only rearwards pressure against a spring to engage (see pages 20, 23, 24, 31, 32, 35, 37, 41, 44, 47, 49, 50, 55). A single front wheel prevented use of this type of starting arrangement and the Wallis Cub Junior (page 25) was started by a lever, giving an impulse similar to that from a motor-cycle kick starter. John Deere tractors usually started from a turn on the flywheel and transversely mounted engines were started from the side of the tractor. Cranking an engine was not difficult for the experienced operator, but tractors were temperamental and the correct set-up was essential. The Fordson was notoriously difficult (see verse on page 48).

A Platypus range of crawlers appeared in 1952 and the company developed world-wide, with premises at Basildon in Essex, before Howard died in 1971.

Engine: Four cylinder vertical
22–27hp at 1200r/min

Gears: Five forward, one reverse
Dimensions: 8ft 4in (2.54m) long,
12ft (3.66m) with rotary hoe
3ft 2in (0.98m) wide (Sugar-cane model)
4ft 0in (1.22m) wide
(Field/orchard model)
Turning circle: 14ft (4.27m) diameter
(Sugar-cane model)
16ft (4.88m) diameter
(Field/orchard model)
Weight: 1 ton 7.5cwt (1.40 tonnes) on steel wheels
1 ton 12.5cwt (1.65 tonnes) on rubber tyres
Price: Model not available in UK

CASE MODEL L · 1930

The claims made after 1918 for various American tractors are remarkably similar and there seems to have been a lot of agreement on what were desirable design elements. The transversely mounted engine was recognised as efficient as it eliminated the need for bevel gears, generally thought to waste power. Economical burning of kerosene was claimed, often in a 'valve in head motor'. Cleanliness of working parts and efficient lubrication seem obvious features and, judging by the frequent reference to ignition systems, starting must have been a regular problem – it was, anyway, usually done with petrol then switching to paraffin.

The J. I. Case Threshing Machine Company of Racine, Winsconsin advised the farmer to visit his local tractor dealer, not to consider price alone (perhaps a sly reference to the Fordson competition) and to have 'a Case tractor on hand all set to go when the frost gets out of the ground'.

Originally a steam-traction-engine-maker, the Case company has produced successful internal-combustion-engined machines from 1911. Three types were offered in 1920, when Case won first prize in the up-to-24hp class at the Lincoln trials.

By 1929 when the C and L models appeared, the

Case, Model L
I am frequently surprised by the never-ending supply of paint- ▶ *able subjects, often found by accident and not always in obvious rural places. I stopped one misty, autumn morning to use a phone-box on the outskirts of Stockton-on-Tees and saw this wonderful combination of verticals and horizontals as I looked out of the kiosk window waiting for a connection. Actually, it's the grounds of a Community Centre; the colours caused by the accelerated effects of distance due to mistiness were enough to make me forget the substance of my telephone call for the moment. These distance effects are seen as a blueing or greying of objects such as trees and hills – a lightening of tone, a loss of colour and a lessening of contrast – caused in normal atmospheric conditions by the accumulation of particles in the air between the eye and the object. This is called aerial perspective and is amplified greatly by mist, to the point where it happens instantly in fog.*

As the chestnut trees have started to turn, the lightest tones are bright ochre yellow and colour moves through the whole range of greens to turquoise and on to complementary mauve in some shadows. The darkest shadows on the nearer trees are dark enough to be dismissed as black, but I seldom use black paint in landscape, preferring mixtures tending towards cool (blue) and warm (brown), leaving the use of black for details of the tractor that are 'coloured' black and to give extra tonal contrast to pull the subject machine out towards the observer.

J.H. APPLEYARD
8/79

Emerson Brantingham Company had been absorbed, the name was simplified to the J. I. Case Company and engines were mounted fore and aft. Despite the abandonment of previous design principles, good performance in the World Tractor Trials in 1930 built a sound reputation for Case tractors in Britain.

These 1930 trials were held at Wallingford, Berkshire and lasted for seven weeks through June and July. Thirty-three tractors took part and each was required to do at least 24 hours' work including tests for drawbar, belt, ploughing and cultivating. The following models took part: Austin (French); Case C; Case L; Fordson (Irish); IH Farmall; IH 10/20; Lanz Bulldog; Massey-Harris; Peterbro; Rushton (the Fordson and Peterbro broke down).

Engine: Four cylinder vertical, 26–40hp
Gears: Three forward, one reverse
Dimensions: 10ft 6in (3.20m) long,
                5ft 8in (1.73m) wide
Turning circle: 21ft 8in (6.60m) diameter
Weight: 2 ton 10cwt (2.54 tonnes)
Price in 1929: £370 (US $1,798)

CASE MODEL R · *1939*

The Case Company has a long association with farm machinery, its foundation going as far back as 1842 when Jerome Increase Case set up in Rochester, Wisconsin, to build threshers. Two years later he moved to Racine, as he was unable to obtain water rights for his factory in Rochester, and by 1848 the firm was Racine's leading industry and largest employer. The eagle trademark, 'Old Abe', which was used from 1865, was based upon a mascot of the Wisconsin Regiment in the Civil War. Four years later they built their first steam engine and by 1892 had produced a 'gas' tractor.

The Case Model R was a versatile, reliable tractor of which large numbers were imported from the USA. The grey model with square-cut radiator was produced for four years until a radical updating in 1939, when the colour was changed to the more familiar Case orange (officially known as Flambeau red), and a rounded, more streamlined front protected the radiator. Standard equipment did not include power take-off, which when fitted ran at 543r/min for binding, mowing, etc.

A letter of 1 November 1938 from the engine firm of Amanco, the British agent for Case tractors, draws attention to new R and RC models, updated to include electric starting, a redesigned engine with a thermostat and improved lubrication. From 20 December 1937, Amanco ran their business from an opulent-sounding address in the Palace of Industry at Wembley.

Chain drive was recommended as putting 'more power to work', spreading the load 'over 60 teeth instead of 4 to 8 gear teeth'. Previous experience with models L and C suggested that few adjustments were needed and the chains probably outlasted the remainder of the tractor.

Between 1967 and 1970, Tenneco Inc of Houston, Texas, gained control, Case becoming finally a wholly owned subsidiary – the 'Old Abe' trademark disappeared in 1969 in favour of what is described as 'a new, more dynamic corporate symbol'.

Engine: Four cylinder vertical, Waukesha/
             Continental
             15–20hp at 1425r/min
Gears: Three forward, one reverse
Dimensions: 8ft (2.44m) long,
                4ft 6in (1.37m) wide
Turning circle: 20ft (6.10m) diameter
Weight: 1 ton 17cwt (1.88 tonnes)
Price in 1938: £225 (US $1,100)

I have to admit that Northcountry villages are generally not ▶ *pretty in the sense that the word has when describing the softer lines of say Sussex or Herefordshire. The more rugged character of limestone or millstone grit shows in this West Riding village, used by Yorkshire Television as the setting for parts of the series Emmerdale Farm – the farm itself is some miles away in Wharfedale. The row of cottages is called Demdyke Row in the television series and has had some windows replaced, but the simple rural versions of the Venetian window to the left of the tree impart an air of elegance to the unadorned indigenous ashlar style. I shall not give the village's name as too much curiosity proves inconvenient for the filmmakers (January 1986: this is now an 'open secret') – the series had to be moved from the village of Arncliffe in Littondale as sightseers became a burden to both villagers and television crews (see page 16).*

The grey stone walls, firmly patterned with mortar joints, and the green in front of the cottages provide an excellent foil for the late, orange version of the R model Case. Trees, behind and in front of the houses, help to soften the hard lines of the stone buildings, but here the effect is accidental and natural, unlike the over-conscious plantings of city-centres and shopping precincts in a vain attempt to humanise concrete and steel.

The styled tractor with its near art deco sunburst radiator and streamlined mascot is short and tubby, but looks rugged and efficient. In shadow and reflection the warm orange colour responds well to the cool light of this October day.

KIRKLEY HALL

J.H.APPLEYARD
9/79

OLIVER MODEL 70 · 1937

In 1929 the Oliver Chilled Plow Works became the Oliver Farm Equipment Company after a merger including the Hart-Parr Company, one of the early pioneers of tractor production. Oliver Hart-Parr tractors were vertical engined and continued under the joint name until 1937, when Hart-Parr was dropped.

The 70 model was introduced in 1935, an early example of a compact six-cylinder tractor, sporting electric self-starting, appearing mainly in the rowcrop version. Its styling was advanced, the radiator being

protected by curved sheet-metal and a dummy grille; a driver's cab was offered by the company in the late thirties. Other models and revisions followed regularly. The largest pre-war Oliver was the four-cylinder 90 in 1937, later offering four-wheel drive, which was also used on the 80 Standard. 1939 saw a revised 70 with six forward gears and the 70 KD, designed to run on paraffin (70 was the octane rating of the petrol fuel recommended). Also just before the war came the 60, with four cylinders and four forward gears with an extra cog added later. In 1944 Olivers merged with the Cleveland Tractor Company to form the Oliver Corporation, the name Cletrac being discontinued from

It is often difficult to rationalise in words the processes by which ▶ a painting comes about. The old excuse that if you could or would put it into words, you wouldn't paint it, is a spurious simplification, a 'cop-out'.

First, I am required to produce an accurate portrait of a vehicle which exists because someone has put a great deal of time, skill and knowledge (not to mention money) into preserving or restoring it. He is certain to be a very well-informed critic of my work. One of the artist's main difficulties is that he is the servant of two masters; he has to deliver the goods and be true to his vision of the subject. As the tractor is the main subject, the painting cannot be merely a landscape containing a tractor – the picture of the Farmall F12 goes to the opposite extreme, justified in my opinion by the complicated mechanisms of the tractor, relieved only by the level green fields beside the River Tees, west of Darlington. Despite the apparent simplicity of the composition, this was one of the more labour-intensive pictures and quite dynamic, although the subject is static, as so many angles and changes of tone are involved. In terms of colour, the interest lies in the wide variety of reds found in the tractor's shiny paintwork – compare underneath the near mudguard with the inner area of the farside guard – and the grey implement and its controls.

J.H.APPLEYARD
3/80

◀ *While the tractor has to take up a large proportion of the picture space, there must be enough setting to make the location believable. In the case of the Oliver, the open barn of a farm near where I live provided the sense of scale, the small hut fills a space to the right of the subject and the dark trees silhouetted against the evening sky provide space. The low sun casts long shadows with interlocking shapes and emphasises the form of the tractor wheels, tyres and engine details – other shadows from the steering wheel on the far mudguard and on the inset radiator grille serve to create a feeling of depth. Red wheels give a colour bonus by way of contrast to an otherwise limited range of colour.*

this date. Further changes followed as the White Motor Corporation bought out Olivers in 1960 and large Oliver tractors became known as Whites. The smaller Olivers were made in Europe by David Brown and Fiat, but the effective import of Olivers to Britain came to an end with the lapse of the lend-lease agreement with the USA at the end of the Second World War.

Engine: Six cylinder, OHV, petrol (70 KD – paraffin)
        23–28hp
Gears: Six forward, one reverse
Dimensions: 12ft (3.66m) long,
        6ft 6in (1.98m) wide
Turning circle: Unobtainable
Weight: 1 ton 11cwt (1.57 tonnes)
Price: Unobtainable

INTERNATIONAL HARVESTER FARMALL F12 · 1937

The successors of the F series, in 1939, were among the American tractors which made a substantial contribution to the British war effort between 1939 and 1945. Designated A, B, H and M, the smallest, A, was petrol powered, the H started on petrol, then switched to paraffin, while the 37hp M offered a choice of petrol or petrol/paraffin. From 1949, following International Harvester's establishment of overseas plants, a 36–39hp version of the M, called the BM, was made at Doncaster and also at the old Jowett car works in Idle, Bradford, Yorkshire.

The Farmall range first appeared in 1922, commercial production starting in 1924, and provided a series of versatile, general-purpose tractors with adjustable track, high ground clearance and centre steering, which permitted a tricycle layout to be adopted for rowcrop work. By 1938 over 400,000 had been produced, including the F30 model, introduced in 1931

and similar to International's W30 model. Pneumatic tyres were offered as an option and the Farmall range was painted grey with red wheels until November 1936, when all-over red was adopted.

Engine: Four cylinders
        10–15hp at 1400r/min (one furrow)
Gears: Three forward, one reverse
Dimensions: 11ft 6in (3.50m) long,
        6ft (1.83m) wide
Turning circle: 14ft (4.3m) diameter
Weight: 1 ton 4cwt (1.22 tonnes)
Price in 1936: £185 (US $919)

FORDSON MODEL N · 1929-45

The successor to the Model F was made in Cork, Ireland, for the first three years of its sixteen-year production run; Irish figures reached 38,600 despite the severe effects of the slump after 1930. From 1932 production was centred at the new Ford works at Dagenham and in 1938 the colour was changed from blue to orange. A further change to green took place in 1940. Early models had rear mudguards extended down into bulges (doubling as small tool-boxes) which were intended to prevent the tractor rearing backwards.

The Model N had no hydraulics, but efficient magneto ignition and an easy-starting device made it a very popular tractor; it seems unnecessary to mention its favourable price. Pneumatic tyres were a regular option to steel wheels and 'Roadless' half-track conversions were common, especially in use in great numbers by the forces in the Second World War. It was also a popular base for experiments with alternative fuels to save precious imports.

Despite severe competition from the Ford-Ferguson, which was made in Ford's own American plant from 1939, the Model N consolidated the Ford grip on the market as a cheap, general-purpose tractor. Regardless of the lack of hydraulics, there were many experimen-

The setting here is the woodyard of the hall at Marske in ▶ *Swaledale, between Richmond and Reeth, deep in James Herriot's beloved Yorkshire Dales country. Just before our visit a BBC television crew had been filming in the grounds of the hall for what I understood was to be a Christmas special (BBC Christmas 1983) based upon further episodes from the famous vet's experiences. I hope they were more careful than I was, for while concentrating upon focus for one of my photographs, I almost backed into the sawpit behind the arches of the building on the left. In such a pit the unfortunate bottom sawyer laboured at the lower end of a two-handed saw in near-darkness and a shower of sawdust.*

The dominant themes are sunlight and green, with contrasts of shadow and pink. Most of the wall areas are in direct or oblique sunlight, picking up the rough stone textures and strong shadow emphasises the form of the Fordson's crankcase and repeats the patterns of the wheel spokes. Variations on the contrasting pink are produced by the faded red-brown paint of the shed doors, the front wheel rims and the clump of bright magenta rosebay willow-herb – the latter a neglected natural beauty, notorious for its invasive, wind-blown gossamer seeds. The yard, with its air of elegant neglect (many outlying farm buildings display more neglect than elegance), overgrown with grass whose cushioning effect makes possible the use of steel wheels with spade lugs, is typical of the places where tractors of this vintage can be uncovered, derelict and dirty, but often surprisingly complete unless cannibalised for spares.

JHAPPLEYARD
9/83

tal, mechanically driven, direct mounted implements, and a variety of adaptations embracing almost every type of powered, wheeled transport from railway locomotives and shunters to tow-trucks for removal vans and the more obvious dumpers and earth-movers.

Engine: Four cylinder, 27.28hp at 1100r/min
Gears: Four forward, one reverse
Dimensions: 9ft (2.74m) long,
          5ft 6in (1.68m) wide
Turning circle: Unobtainable
Weight: 2 tons 3cwt (2.18 tonnes)
Price: £178 10s 0d (US $720) in 1945;
     £155 (US $753) in 1929

THE GREEN MACHINE

Short and squat a green machine
In the 40s came on the scene,
A standard Fordson stark and bare
To cause a lad to stand and stare.
A modern machine with a mind of its own
That at certain times was known
To take control, the memory lingers
Of rattling of many fingers
By the handle or the steering wheel.
And in my mind I still can feel
A fusion of machine and man
For time would settle a plan
Come to know that mind of its own
Hear the gearbox whine and groan
To sound at night that more content
Especially when off home you went.
Whilst on a morning cold and damp
It moved as if it suffered cramp
Had to be coaxed each time to start.
To hear that engine like a heart
The smell of paraffin in the air
Sit on a seat cold and bare
Wait for the clutch to disengage,
I drove a Fordson at an early age.

© Jim Anderson

MASSEY-HARRIS 4×4 · *1930-36*

The four-wheel-drive idea had long been seen as a way to achieve maximum traction and many examples were offered by a number of smaller companies; in addition to the 'Iron Horse' from the Samson Division of General Motors, 4×4 tractors from US companies in the twenties were the Wilson, the Wizard and Fitch and Storey's pivot-steering model (page 36 for reference to Mr Storey). But this Massey-Harris 4×4 general-purpose model was the first practical four-wheel drive and the first of their own designs, previous models having been based upon Wallis tractors (page 26) and the Parrett.

Good ground clearance of 30in (0.75m) was achieved by applying the drive to reduction gears at the top of the rear wheels and a pivoting rear axle ensured good adhesion on uneven ground. The general layout was advanced for its time, the forward engine mounting being the standard practice for modern high-horse-power tractors. Altogether this was a successful early example of four-wheel drive, although Massey-Harris introduced a more conventional model, the 25/40 related to the Wallis, in 1933.

Engine: Four cylinder vertical (Hercules)
          25hp at 1000r/min
Gears: Two forward, one reverse
Dimensions: 10ft (3.05m) long,
          4ft to 6ft 4in (1.22m to 1.93m) wide
Turning circle: 6ft (1.83m) diameter (with use of brakes)
Weight: 2 tons 10cwt (2.54 tonnes)
Price: £450 (US $2,205)

I had not thought of writing notes on these tractor paintings ▶ until David St John Thomas suggested it and sent me a copy of The Railway Paintings of Don Breckon, *published by David & Charles. The idea of trying to explain the development of my work had seemed overly self-indulgent, but I saw that such explanations could reveal to the average reader processes not readily understood and justify the pictures' subject matter as well as clarifying the ways in which the pictures are composed.*

Consequently, all the notes until now have been written about completed paintings. In this instance I intend to write about my intentions and aims as the Massey-Harris painting is not yet started (December 1983). This is not to contradict previous statements that firm plans seldom precede starting work on a picture – I can indicate only general planning; there is no sense in which I ever completely visualise the form that a picture will take.

I feel that the tractor does not require much by way of compensating shapes in the setting and expect to use the solid dark lines of the hay-barn to silhouette the broken red outline of the tractor. Incidentally, this rural setting on the fringe of a now-bypassed village is close to the roaring traffic of the A66 road between Middlesbrough and Darlington, before joining the motorway and becoming the trans-Pennine road from Scotch Corner to Penrith and beyond.

The dull green colour of the field provides a perfect complementary contrast with the tractor's colour, and as the day I went sketching had been showery, I shall try rain-puddles on the track to give interesting reflections of the tractor's shapes and to introduce a lighter, sky-coloured patch in the solid foreground. Effects of space and distance will rely mainly upon the perspective of the track and the few small details of the fence and gate behind the tractor, but something, such as a stand for milk-churns, leading into the picture from the bottom edge would help to push the machine back into the picture-space. We shall see – where can I find a milk-churn?

J.H. APPLEYARD
1/84

ALLIS-CHALMERS MODEL U · 1949

Allis-Chalmers are best remembered as the firm which adopted the newly developed Firestone pneumatic tractor tyre in 1932. Its herringbone tread gave a degree of grip on the land comparable with that from metal wheels with cleats or spade lugs. On the road, the advantages were obvious, particularly to other road users and 67.87 miles/hour (109.2km/hour) was

achieved as a publicity stunt by a specially geared-up Model U. This was probably a world speed record for tractors and may still stand, although there is no mention of it in the *Guinness Book of Records*.

The Model U started as the United tractor, manufactured and sold by a consortium of American agricultural machinery companies. This venture of 1929 did not thrive, but Allis-Chalmers continued to make the design (U for United) using their own engine. It was particularly popular with agricultural contractors for threshing.

The later Model M was virtually a tracked Model U,

It is very difficult for the artist to avoid becoming obsessive – not with ecstatic reactions to sunsets or boring examinations of perspective and certainly not externalised for public display, but quietly observing and recording and building up a lifetime of experience of the nature of things. I regard my work as much a part of living as breathing and so there is nothing extraordinary in the fact that I do not stop drawing simply because I am on holiday. In fact, new places are full of fresh stimuli and challenging because of their unfamiliarity.

The setting for this John Deere rowcrop machine is based upon observations made near Tenby, in what used to be Pembrokeshire. Above the town, reached by a rough, narrow path, there is a viewpoint provided with seats and a chart, giving directions and distances to points of interest visible from this hillside. The area is dedicated to some late local worthy whose favourite viewpoint this was; but climb a little farther, away from Tenby and there's a much less organised view, over Saundersfoot and towards Pendine Sands, which I have used for this picture.

The basic theme is light and shade, expressed as warm and cold, but with the complication that blue has also to be used for the sea and distance in the far cliffs and fields. The contrast between the tractor's green paint and the fields to its left is subtle, while stronger differences are presented by the brightly lit cornfield (spoilt by patches of sorrel) and the tractor's lemon wheels. The lighting is capricious, with sunshine breaking through the trees 'out of shot' to the left and dappling the fence-posts and just touching the tractor's upper parts, leaving the foreground details only partly lit and cool in colour.

J.H.APPLEYARD
7/80

The hot bright orange of the Allis-Chalmers range of tractors is ideally set off by the blues of shadows and distant hills. Other cool colours appear in both the highlights which reflect the blue of the sky and the shadows on the tractor body and wheels. The machine is shown here on steel wheels and with some of the rear-wheel cleats removed in preparation for use on hard surfaces – metal cleats could inflict such damage on road surfaces that their use was soon outlawed. Imagine the effect of their lumpy progress on tarmacadam upon the operator's spine! The landscape is just outside Hovingham in North Yorkshire, the early home of HRH The Duchess of Kent and seat of her family, the Worsleys.

possibly designated for the Monarch Tractor Company, who were makers of crawlers, taken over by Allis-Chalmers in 1928.

Engine: Four cylinder vertical
40hp at 1200r/min
Gears: Four forward, one reverse
Dimensions: 9ft 2in (2.79m) long,
5ft (1.52m) wide
Turning circle: 15ft 6in (4.72m) diameter
Weight: 2 tons 1cwt (2.08 tonnes)
Price: £462 10s 0d (US $1,702)

JOHN DEERE MODEL AN · 1948

The first true John Deere tractor, the Model D, appeared in 1923 to replace the Waterloo Boy (Overtime) (page 29) which had worn the characteristic yellow and green colours since the take-over of the Waterloo Gasoline Engine Company in 1919. The 10/20 General Purpose model in 1928 was the first popular tractor with a mechanical implement lift. After a considerable contribution to the British war effort by the Overtime from 1917, John Deere tractors were not actively marketed in Britain until the thirties, when the one-furrow Model B was imported in 1935. They then virtually disappeared until 1962, to be seen in regular use up to the present in areas where a local dealership ensures that a wide range of machines, including combines, is to be found on farms.

The whole range was restyled in 1938 with sheet-metal fronts. The 1938 Model B was a four-wheeler with close-set front wheels, the BN being a three-wheeler of 16–20hp. The first John Deere with a diesel engine was the largest offered, the Model R in 1940 – larger tractors being needed to offset reduced manpower on farms of ever-increasing size. As became more common, after the Second World War, power take-off and pulley were standard fittings.

Engine: Two cylinder horizontal
31–37hp at 975r/min
Gears: Four forward, one reverse
Dimensions: 11ft 1in (3.38m) long,
6ft 11in (2.11m) wide
Turning circle: 16ft 8in (5.1m) diameter (Model A)
Weight: 2 tons (2 tonnes) approx.;
1 ton 14cwt (1.73 tonnes), Model A
Price in 1942: £410 (US $1,763)

LANZ BULLDOG MODEL 22/38 · 1935

Until the introduction of the full diesel engine in 1953, the Lanz tractor followed the continental fashion for the hot-bulb engine, running on heavy oil. Starting was unorthodox; the steering wheel and column could be withdrawn and inserted into the centre of the flywheel. Meanwhile, the cylinder head was heated by a blow-lamp, and when it was hot enough turning the steering wheel then had the same effect as using the normal crank-handle and the compressed mixture was brought into contact with the 'hot-spot', causing the fuel to burn and starting the normal four-stroke cycle. Subsequent explosions kept the hot-bulb at a sufficiently high temperature to continue igniting the compressed mixture in the cylinder. Although no spark plug was used, this is different from the true diesel principle where ignition heat is produced by the high compression of the mixture. Solid rubber tyres were offered as an optional extra on this design from H. Lanz of Mannheim, which was imported into Britain by the Locomobile Engineering Company of London and Manns of Saxham in Suffolk.

During the forties and fifties, continental tractors continued to use simple, single-cylinder designs, principally employed for towing and haymaking with attached implements.

Engine: Single cylinder, horizontal
15–30hp (drawbar pull 2,000lb)
Gears: Three forward, one reverse
Dimensions: 9ft (2.74m) long,
5ft (1.52m) wide
Turning circle:
Weight: 2 tons (2 tonnes) approx.
Price: £325 (US $1,592)

This is another tractor using the popular green and red colour ▶ *scheme, but technically unusual and displaying an interestingly varied broken outline which helps to bring the subject forward from the landscape skyline. Gates figure frequently in the paintings because they suggest the sort of place where a tractor would be halted on its way to or from work and provide strong horizontals to contrast with the vertical lines of the machine. I often feel that my tractors are too clean to have been working in the settings portrayed, but amounts of muddy earth clinging to the wheels would blur or obscure the detail which is one of my main considerations (but, see page 64). A farmer conscientious enough to clean his tractor wheels in this fashion would be popular with any motorist who has had his windscreen spattered while following a ploughing rig along a country lane or contrived to stay out of the ditch on a mud-covered road. The landscape is farming country near the market town of Stokesley in North Yorkshire, in the shadow of the Cleveland Hills.*

FERGUSON-BROWN · 1936-39

As a dealer, Harry Ferguson had run an agency in his native Ulster for the Overtime tractor (page 29). But as early as the First World war and again in the twenties he had experimented with ideas to overcome two of the major shortcomings of the lightweight tractor. These were: (1) insufficient traction at the drive wheels when pulling a heavily working implement; (2) the tendency of the tractor's front wheels to lift as a reaction to the implement's drag – the wheely effects of the stunt motorcyclist and the speedway machine, when the clutch is let in, are not welcome in tractors. Others were aware of the danger; the early Fordson N (page 47) had low rear mudguards with a bulge intended to stop the tractor rearing backwards.

Henry Ford was interested in Ferguson's ideas and the British firms of Rushton (backed originally by AEC in the late 20s) and Ransomes and Rapier proposed to use the Ferguson system. However, he did not wish to work for Ford and the British companies were not powerful enough to fulfil Ferguson's vision of the importance of his system.

By 1933 the 'Black Ferguson' prototype, with some parts made by David Brown of Huddersfield, was demonstrating the integrated tractor and implement system with great success, and in 1936 Browns continued the collaboration when they undertook production of the model, unofficially known as the 'Ferguson-Brown'. There was a formal merger of D.B. Tractors and H. Ferguson Ltd in June 1937 to form Ferguson-Brown Tractors Ltd. The Hercules engine of the prototype was replaced by a Coventry-Climax unit at first and later models used an engine developed by David Brown. But, including implements, the price was twice that of a Fordson and the tractor sold very slowly. For many farmers it was also too small and a larger version was proposed by David Brown and opposed by Harry Ferguson. This disagreement and the general lack of success made Ferguson look elsewhere for backing and production; a demonstration to Ford at the Dearborn farm convinced both of them that the best prospect lay in co-operation. Ford's production methods would drastically reduce the cost and Ford saw enough advantage in access to Ferguson's ideas to outweigh the fact that the Ford with Ferguson system would provide strong competition for Ford's own N Model (page 47). The first Ford-Fergusons appeared in June 1939, one month after David Brown's VAK (page 60) model which used many Ferguson features, but had a 35hp engine and built-in power take-off.

One drawback of mounting an implement, especially a plough, directly to the tractor was that the implement dug in if the tractor's front wheels went over a bump and surfaced if they went into a hollow. To counteract this, Ferguson developed his draught-control system. A pre-set depth could be maintained by a unit connected to the hydraulic pump in the lifting linkage, with manual override for variable soil conditions, but this was not a standard feature until the British-made TE models were introduced.

It is tempting to say that the Ford-built Ferguson looked like the grey Ferguson (page 57) so familiar still on many farms, but this would be like saying that a father is like his son, for the British-made Ferguson in the period after 1946 was a near copy of the machine made by Ford throughout the Second World War.

Engine: Hercules, four cylinder (prototype 1933)
　　　　 Coventry-Climax
　　　　 18–20hp at 1000r/min
Gears: Three forward, one reverse
Dimensions: 8ft 9in (2.67m) long,
　　　　 5ft 1in (1.55m) wide
Turning circle: 21ft (6.4m) diameter
Weight: 17cwt (0.86 tonnes)
Price in 1938: £224 (US $1,095)

I've frequently read statements by novelists and playwrights ▶ that they sit down to write with the barest outline of their plot in mind and that the characters take over – and I've been sceptical about the genuineness of the claim. But having undertaken the task of analysing my approach to these paintings, I see that there is a similarity between writing and painting and understand something of what they mean.

The decision to show two views of the Ferguson-Brown was deliberate – one view to show the three-point linkage attaching a plough and the other to give an impression of the tractor's proportions and to include the traditional radiator with the Ferguson name. Additionally, this gave me the chance to portray the tractor on both steel and rubber. The setting was chosen to provide a light background for the silhouetted rear view and the tractor in the foreground moves from light to shade in three ways: side-lighting on the vehicle itself, cast shadows, and a section merging into the dark hedge and gatepost.

The stubbled field with its fringe of dark trees is once again near Hovingham in North Yorkshire, looking towards the Vale of Pickering. For Yorkshire, this area, and the larger, nearby Vale of York, is relatively flat and intensively farmed. The sky indicates a breezy day of sunny periods and possible showers; bumpy weather among the cumulus clouds for the aeroplanes from the once-crowded airfields of this area, second only to Lincolnshire for bomber stations of the Royal Air Force.

Having made the initial drawing, in pencil on a prepared board surface, I then have to 'fill in' the spaces between the lines. The line drawing is as detailed as possible, each line indicating an intended change of colour or tone. The landscape is then brought to a nearly complete stage before painting of the tractor is started, although tentative touches are made here and there to test the lightness and darkness of the tractor tones in relation to the setting. This is too calculated as it reads, for the painting has already 'taken over'. I am seldom aware of what I would call deliberate thought about what colours and tones are appropriate or how they are achieved by the physical mixture of oil pigments. I suppose it comes down to instinct and experience, although the latter does not take into account the case of highly talented young painters – but then this account is an attempt to explain me at work and certainly we are not all the same.

FERGUSON TE MODELS

See page 54 for an account of the early relationship between Ford and Ferguson.

Ford and Ferguson rubbed along together for seven years, perhaps forced to sink their differences for the common good during the Second World War, but generally with more friction than agreement. Ford made the tractors (9N petrol; 9NAN paraffin; 2N Utility), Ferguson marketed them and would not sell out to Ford and see his name disappear from the product, nor would Ford agree to Ferguson's demand that the tractors should be made in the UK. November 1946 saw the end of what had been known as a gentlemen's agreement between the two and when Ford introduced his 8N model, Ferguson filed a quarter-billion dollar lawsuit claiming damage, conspiracy against him, patent infringement and unfair competition. By 1953 Ford was willing to pay out $9.25 million to be free of the threat of further lawsuits.

From the break with Ford, the familiar grey Ferguson was made at the Standard Motor Company's works at Banner Lane, Coventry and designated Model TE, for Tractor England. By 1952 Perkins diesel conversions were available and in 1953 the amalgamation with Massey-Harris made Massey-Ferguson the world's largest producer of tractors.

In literature for the TE models, the Ferguson was proclaimed to be 'The World's Most-copied Tractor' which betrays Ferguson's understandably defensive attitude and indicates the attractions of his system. The tractor was no longer to be merely a puller of implements and a power source, but a fully integrated system to perform all tasks on the land. An impressive list of twelve 'firsts' includes attention to problems of hydraulics, draught control, traction, overload protection, adjustable widths and many mounted implements. About forty categories of implements were offered, in addition to accessories, all conforming to the stated intention 'to make tractor and implement one single operating unit'. Besides being an inventor with that gift for creative leaps into new areas of experiment, Harry Ferguson was nobody's fool in the field of marketing philosophy and the purchase of a Ferguson tractor guaranteed a ready sale for Ferguson implements. The slogan was 'It's what the implement does that sells the tractor'.

The following section paraphrases the explanation of the Ferguson system given in the dealers' handbook – The Ferguson system utilises three points of attachment to transmit the pulling force of the tractor to the drawn implement. Two bottom links pull the implement and a top link pushes forward and down on the tractor above the rear axle. So the weight of the implement and soil resistance to it add traction-producing weight as needed. The front end of the tractor is kept down by the forward thrust through the top link. A control spring in the top link reacts to compression or expansion caused by the forces on the top link to operate a control valve which regulates depth and gives automatic draught control. Strain on the tractor and implement is avoided in the event of an obstruction by a control spring which releases the implement on impact and transfers weight to the front wheels, causing the rear wheels to spin and a loss of traction.

Various combinations of the assembly of rear wheel discs and rims give adjustment of track from 48in (122cm) to 80in (203cm).

So there are three phases in the Ferguson story. (1) Initial experiments culminating in a 1933 prototype followed by production at David Brown's Huddersfield works, this model was never officially called the Ferguson-Brown (page 54). (2) American production from 1939 to 1946 of over 300,000 Fords with the Ferguson system. (3) From 1946 until the Massey-Harris merger in 1953, the production of well over 300,000 TE models at Coventry. This is the best-known grey Ferguson with its typically American-styled front inherited from the Ford-Ferguson, still to be seen in quite large numbers by even the only casually observant on farms all over the land.

Engine: TE/A/20 petrol, 24–28hp
TE/F/20 diesel, 26hp
TE/D/20 TVO
Gears: Four forward, one reverse
Dimensions: 9ft 7in (2.92m) long,
5ft 4in (1.63m) wide (standard width)
Turning circle: 17ft 6in (5.33m) diameter;
16ft (4.88m), petrol model with use of brakes
Weight: 1 ton 2cwt (1.12 tonnes), approx.;
1 ton 5cwt (1.27 tonnes), diesel
Price: £325 (TE/A/20) (US $1,310)

The grey Fergy, possibly the most familiar tractor of all, was ▶ portrayed against brick buildings to set off its no-nonsense, functional paintwork. Because there were and are so many of this model to be seen, I included a second example with modifications such as the vertical exhaust, headlamp and home-made roll-bar.

The farm buildings, now demolished and covered by a light industrial site, were on the outskirts of Darlington in a run-down area of mixed terrace houses and industry. I believe that the farm considerably pre-dated its Victorian surroundings, though not a particularly distinguished piece of architecture, and it was interesting to speculate on its surroundings in its heyday. The ploughed land, trees and distant landscape were added from another location to give additional space and colour contrast with the warm-toned brick buildings. The atmosphere is intended to be that of an early spring day – the farm still has fires burning – crisp and breezy and with little sign of new life in the ground. The land close to the farm, which most of us would call the garden, is rough, overgrown and neglected, all the owner's energies being channelled into work on the fields.

RANSOMES MG/ TRUSTY STEED/OTA/ SINGER MONARCH

In the postwar years, when Britain still lived with rationing and made wry jokes about 'Utility' goods, a number of companies not normally associated with tractors brought out lightweight machines intended for use on small acreages, lighter jobs and in horticulture. 'You wouldn't use a hammer to crack an egg' stated an advertisement for the Singer Monarch in 1955 – it was reviewed in 1951 as a four-wheeled companion for the three-wheeled OTA, both then being offered by Oak Tree Appliances of Coventry and capable of interchanging all implements, such as a single-furrow, general-purpose plough and disc harrows.

Tractors (London) Ltd of Barnet made two-wheel, market-garden cultivators throughout the thirties and the Second World war. The 'Steed' model was a four-wheel, ride-on model intended as a cheap, light tractor using various proprietary engines and pressed-steel car wheels. A centre-mounted tool-bar gave good vision of work in hand, and rear-wheel steering ensured a short turning range. During the twenties this market was provided for by adaptations of the Ford Model T, Morris Cowley and Austin 10/12 cars. From 1936 Ransomes, Sims and Jefferies of Ipswich made a 9.5cwt crawler for market gardeners. With one forward and one reverse gear, the MG2 sold for £133 (US $661) in the thirties and was used extensively by the services during the Second World War, mainly as a towing vehicle for bomb trolleys. By 1960, when production ended, the MG6 model was the result of 'progressive development and improvement' and could be fitted with a hydraulic implement lift as well as a 48in (122cm) bulldozing blade.

Ransomes MG Series – 1936–59
Engine: Single cylinder petrol or TVO
7bhp
Gears: Three forward, three reverse
Dimensions: 6ft 8.5in (2.04m) long,
3ft 2in (0.96m) wide
Turning circle: Steering by lever-operated brakes
Weight: 13cwt (0.66 tonnes)
Price: £335 (US $935) in 1955;
basic model £133 (US $661) in 1936

Trusty Steed – 1950
Engine: Single cylinder, air-cooled Jap or Norton
Rated 4hp, claimed to develop 14hp
Gears: Three forward
Dimensions: 9ft 5in (2.87m) long,
3ft 4in (1.02m) wide
Turning circle: 'Practically within wheelbase'
(two-wheel model)
Weight: 5.5–6cwt (0.28–0.30 tonnes), two wheel model
Price in 1950s: £206 (US $577)

OTA – 1949–52
Engine: Four cylinder, Ford 10 industrial
17hp at 2000r/min
Gears: Three forward, one reverse
Dimensions: Unobtainable
Turning circle: Unobtainable
Weight: 15cwt (0.76 tonnes)
Price in early 1950s: From £245 to £269 (US $686 to
$753)

Singer Monarch – 1952
Engine: Four cylinder, Ford 10 industrial
17hp at 2000r/min
18.5hp (1955)
Gears: Six forward, two reverse
Dimensions: Unobtainable
Turning circle: 7ft 10in (2.39m) diameter
Weight: 13cwt (0.66 tonnes) approx
Price: £297 10s (US $833), 1951/52;
£279 (US $778), 1955 – Singer

This is a very well-equipped market garden, which does not ▶ *show much marque loyalty, although both the OTA and the Monarch were made in Coventry by Oak Tree Appliances. Of course, the painting is devised to exhibit a number of small tractors which were popular on small farms and for light agricultural and horticultural tasks in the late thirties and in the years immediately after the Second World War. Most notable are the different shapes evolved to perform approximately the same tasks – it is this variety that makes it possible to group them closely without too much repetition. The Monarch, although small, is of conventional layout; the three-wheeled OTA only marginally less so; but both the Trusty and the Ransomes MG are eccentric designs, the latter a miniature crawler and the other an obvious assembly job using a number of proprietary parts and owing little to contemporary tractor design. Its open construction and prominent controls allowed me to create space by superimposing foreground details upon the horizontal features of the setting, an effect enhanced by the sturdy gateposts and the two open gates.*

I had had the form of the painting in mind for some months since I called at the nursery shown in the early summer for some dahlia plants. The actual greenhouses are modern aluminium structures, but I have used the bolder patterns of the older wood-framed type which are more in keeping with the period when these machines were active. The possibly monotonous roofline of these buildings is relieved by introducing two trees, which also show through the glazing, implying more space, and the chimneys of the heating plant. Observing the painting you are looking north, so the light is that of a calm, late summer afternoon, bright enough to cast shadows, but somewhat diffused and without much warmth.

J.H.APPLEYARD
12/83

DAVID BROWN
MODEL VAK IC · 1947

The 100th Royal Show at Windsor was the occasion for the launch of the VAK I model in 1939. It was a more powerful version of the Ferguson-Brown page 54 with built-in power take-off, a feature which David Brown had wished to include on the model they made for Ferguson, but which had been available only as a bolt-on accessory. Compared with the traditional style of the Ferguson-Brown with exposed radiator, this model had a styled front end and was built on a frame rather than of full unit-construction. Coming from the new works at Meltham, it became a familiar towing tractor for both the Royal Navy and the Royal Air Force; 200 crawler versions were provided for the Ministry of Supply, and the magazine *Flight* in February 1940 was recommending the tracklayer as ideal for aerodrome use. At the end of the war, a forward belt pulley was added to produce the Threshing Model.

The design was updated as the VAK IA in 1945 and two years later the IC 'Cropmaster' appeared, with the 'Taskmaster' as the equivalent industrial model.

In 1956 there was a drastic redesign of previously traditional layout and the result was the DI model. A two-cylinder, air-cooled, diesel engine mounted at the rear gave 12bhp, with four forward gears and one reverse, the lifting mechanism being pneumatic. A mid-mounted tool-bar gave the operator an excellent view for planting and hoeing and the tractor could, of course, be used for conventional towing. Further models included the 900 diesel, with 40bhp and six forward gears. TVO and petrol versions were available. Since 1972 David Brown has been owned by Tenneco (Case).

Engine: Four cylinder, 2.5 litre (Dorman)
25–35bhp at 600–2000r/min
Gears: Four forward, one reverse;
six forward, two reverse option
Dimensions: 9ft 3.75in (2.84m) long,
5ft 0.625in (1.54m) wide
Turning circle: 21ft (6.40m) diameter
Weight: 1 ton 9.5cwt (1.50 tonnes)
Price in 1951: £477 (US $1,336) (Cropmaster);
£863 (US $2,416) (Trackmaster)

David Brown VAK IC Cropmaster and Trackmaster
One of the many things you learn as an aircrew navigator with ▶ the Royal Air Force is an appreciation of weather and the clues to its prediction indicated by the clouds. The ability to observe and report on visibility, cloud base and type and other items such as the surface of the sea was used, not only as a factor in self-preservation, but as a useful supplement to reports received from other sources of meteorological information. I have remained acutely aware of skies and the 'gen' to be gained from their study, as well as being intrigued by their infinite patterns and beauty. Depicted here is changeable weather over the North Yorkshire Moors, less wet generally than that caused by the mountains of the Lake District, but giving diffused, cool light with occasional flashes of sun and shadows flitting across the contours of the hills. Breezes keep the clouds moving briskly and the air is clear, showing two prominent landmarks – Roseberry Topping to the left and Captain Cook's monument, in memory of one of Cleveland's distinguished sons. Both are accessible to the public and the area has mellowed scars from the once-thriving ironstone mining industry.

I have avoided the obvious ploy of showing the two versions of the DB VAKI tractor facing opposite ways to achieve balance and have tried to counter the strong movement from top right to bottom left by emphasising other diagonals, especially in the heavy sky, and with firm, stabilising horizontals in the gate, fence, pond and the dark hills.

Nuffield M4 Universal · 1948

When Allied forces landed in Europe on D-Day, among the items of equipment which were captured were a number of 10hp Light Utility trucks which the Germans had used since they were left behind at Dunkirk. They had been part of the British Expeditionary Force and were made by Nuffield Mechanisations, a company set up in 1936 to make military vehicles, including also an 8hp Utility based upon the pre-war Post Office van and the 'Gutty/Mudlark', which developed into the Austin Champ. After the end of the war, production capacity was underused and the Nuffield tractor was developed to replace lost work, appearing in 1948, the company then being called Morris Motors Ltd (Agricultural Division) of Cowley, Oxford.

A hydraulic lift and power take-off were standard and the tractor was one of the first with five forward gears – six sources of implement power were available. The Nuffield went diesel in the fifties, first with a Perkins P4, then with a BMC four cylinder on the Nuffield Universal DM4 at the quoted price of £610 (US $1,708) basic. The parallel PM4 had a petrol engine and the M4 ran on TVO.

BMC and Leyland merged in 1968 and the company's tractors were called Leyland from December 1969, when the colour was changed from orange to blue.

Engine: 3.4 litre BMC diesel (DM4)
        30hp
        Four cylinders, based on Morris
        Commercial ETA
        33–38hp at 1400 to 2000r/min
Gears: Five forward, one reverse
Dimensions: 10ft 3in (3.12m) long,
            6ft 11in (2.11m) wide
Turning circle: 21ft 8in (6.60m) without brakes
Weight: 2 tons (2.03 tonnes)
Price: £610 (US $1,708) (basic diesel);
       £490 (US $1,372) (TVO);
       £667 10s (US $1,869) in 1950

I start my paintings with a bundle of sketches and photographs which provide the technical details of the particular tractor. Then I search through drawings and photographs of suitable settings, generally choosing one whose colour range gives contrast with the tractor's colour. The foreground space has to be big enough to accommodate a large image of the tractor, which usually takes up two-thirds of the picture's length (36in, 92cm) and up to three-quarters of the height (24in, 61cm), depending upon the tractor's proportions. The painting is seldom conceived as an entity until these two elements are brought together. Then sizes, perspective and tonal ranges have to be adjusted to give the impression of a real machine in an actual setting. The Nuffield M4 was painted as if standing in the farmyard of the Home Farm complex of the North of England Open Air Museum, Beamish, County Durham. The prominent chimney, not actually visible from this angle, shows that the farm was equipped with a steam engine for powering such processes as threshing – the machinery still exists in situ. I am indebted to Mr K. Christensen of the museum for the information that the farm's ground-plan is in the form of a capital E, possibly a reference to the Eden family, the owners of the estate after the original owners, the Shaftos. The Edens apparently habitually wore the top hat (males only, one supposes), referred to in the shape of the chimney's crest, which more practically provides an inspection platform. Beamish Museum, a few miles off the A1(M) in County Durham, near Chester-le-Street, is a unique collection of reconstructed industrial relics, including a working tramway and railway, a pithead and drift-mine, miners' cottages, a full-scale working replica of Locomotion of the Stockton and Darlington railway, a growing town centre which includes a functioning public house and many other items of nostalgia, including the Home Farm portrayed.

The mainly cool stone and slate colours of the buildings are intended to throw forward the red-orange colours of the tractor's immaculate paintwork. Two patches of green, in the hillside and the moss-covered gatepost, give a stronger colour contrast and diffused light comes from a sky of scudding low clouds indicating breezy, changeable weather.

The imaginary registration number (once actually issued in Bootle) is a tribute to J. E. Moffitt of the Hunday Museum, my patron for this and many of the other paintings.

The hens belong to my son-in-law and as they peck around the farmyard, always hungry and curious, they fill the big, empty foreground and give animation and scale. I am sometimes accused of painting a world inhabited only by machines, but not many of my pictures seem to me to cry out for live inhabitants and I find that arrested movement in figures can conflict with the static but live landscape.

JEM983

NUFFIELD

J.H.APPLEYARD 4/83

FIELD MARSHALL
SERIES III · 1949-52

Marshall & Sons Ltd of Britannia Iron Works, Gainsborough, were founded in 1876 and built a reputation as makers of steam engines. The first Marshall tractor, the 15/30, appeared at the 1930 World Tractor Trials. Marshalls followed what is generally called continental practice (see page 52), with simple, two-stroke diesel engines of large bore and stroke (8in × 10.5in, 20.3 × 26.7cm). Starting was with the aid of a glow paper or cigarette end and the tractor was priced just over £300 (US $1,470) in 1930. Other models appeared during the thirties following a policy of continuous development, the most obvious change being in the styled appearance of the 1936 Model M.

Postwar, the Series I was introduced in 1945; two years later the Series II ran until 1949, succeeded by the Series III until 1952, the latter with steering brakes. The IIIA was produced until 1957 and then single-cylinder tractors were discontinued. They were good for stationary duties as they could run indefinitely on full load, using less than one gallon (4.5 litres) of diesel per hour. The IIIA gave 40bhp at 750r/min, its governed maximum and this slow-running engine made the sound which gave the Marshall its nickname of the 'Pom-Pom'. The cylinder head protruded under the radiator and took a cartridge which blew the engine into life, so long as the piston was at the 'top' of its stroke.

Between 1957 and 1960 the last wheeled machines were made using Leyland diesel engines, almost all of these large, powerful machines being exported. The company now concentrates on Track-Marshalls.

Engine: Single cylinder ('semi-diesel') 40hp at 750r/min
Gears: Three forward, one reverse
Dimensions: 8ft 9in (2.67m) long, 6ft 2in (1.88m) wide
Turning circle: 9ft 6in (2.9m) diameter (Series I)
Weight: 2 tons 18cwt (2.95 tonnes)
Price: £597 (US $1,672)

This beautifully kept tractor, over thirty years old, gave me ample scope for expressing my delight in the effects of light reflected in bright paintwork. Every nuance of green can be explored, from near-white to turquoise, so blue that it oversteps the indefinable barrier between blue and green. Reflections such as those of the headlamp, exhaust stack and steering column reveal how good the finish of the panels can be. The tractor's colour is picked up in the sunlit verge where new growth is starting and contrasted with the orange pink of the farm building, typically in this area of north-east England roofed with burnt pantiles. The rendered farm cottages in warm, ochre wash balance the untidy hedge of elder and willow and recede behind the barn with their cooler, purple reds and surrounding vegetation in duller greens. The day is cool and crisp and typical of the weather when the stationary sketcher is grateful for the sun on his back. Cirrus cloud has thickened into a layer of thin stratus which probably means rain before long.

Since starting to write these notes on the paintings, I have been made aware of the need sometimes to combine the detail of the tyre treads with the impression that the tractor has been at work by including mud in the tread recesses. Not too much mud and some juggling with colour and tone seem to overcome my previous fears of loss of detail.

JHAPPLEYARD
3/83